GREENHOUSE GARDENING MADE EASY FOR BEGINNERS

Step-by-Step Instructions for Building and Growing in Your Own Greenhouse

OWEN HAYES

COPYRIGHT

CONTENTS

INTRODUCTION

"In the world of gardening, a greenhouse is not just a structure but a gateway to endless possibilities, where each season blooms regardless of the weather outside."

Have you ever wondered why the sight of lush, green plants can lift your spirits instantly? Is there a magical connection between the color green and our well-being, or perhaps a deeper, more primal bond between humans and nature?

These questions are at the heart of why we garden, and why greenhouse gardening, in particular, holds a special allure. Welcome to a world where you can create your own lush, green oasis, no matter the season or the climate. In this book, I'm not just going to teach you how to grow plants; I'm going to share a passion for cultivating life.

Gardening has been a part of human history for thousands of years, a simple act that connects us to the earth and, in many ways, to ourselves. But why a greenhouse? The answer is simple yet profound: a greenhouse is a sanctuary. It's a place where you can control your environment, coaxing seeds into seedlings and nurturing them into mature plants, regardless of what's happening outside.

It's a haven where you can escape the stress of daily life, immersing yourself in the rhythms of growth and renewal. Imagine stepping into your greenhouse on a chilly winter morning, the air inside warm and humid, the scent of earth and greenery surrounding you. This isn't just gardening; it's a therapeutic experience, a retreat into tranquility.

I remember the first time I walked into a greenhouse. The world outside was a dreary gray, but inside, it was a vibrant, living world. Rows of tomatoes, cucumbers, and herbs thrived in the warmth, oblivious to the cold. The feeling of stepping into that space was like entering a different realm—one filled with potential and promise.

Since then, I've been hooked. The beauty of greenhouse gardening is that it offers a controlled environment where you can experiment and innovate. Whether you're a novice gardener or someone with years of experience, a greenhouse provides the perfect playground for your horticultural creativity.

In my many years of gardening, I've learned that plants are some of the best teachers. They require patience, attention, and care—qualities that, once cultivated, can transform not just your garden, but your life. Greenhouse gardening, in particular, demands a keen understanding of the delicate balance between light, temperature, humidity, and ventilation. It's a science, but also an art.

And like any art, it comes with its own set of challenges and rewards. But don't worry, I'm here to guide you through every step, from the basics to the more advanced techniques. Together, we'll unlock the secrets to creating a thriving greenhouse garden that will provide you with fresh produce and a sense of fulfillment year-round.

One of the most satisfying aspects of greenhouse gardening is the ability to grow what you love, whenever you want. Fancy fresh basil in the dead of winter? How about vine-ripened tomatoes in early spring? A greenhouse makes it all possible. It's your personal Eden, a place where the rules of the outside world don't apply.

But with great power comes great responsibility. A greenhouse is not just a set-it-and-forget-it project. It requires ongoing attention and adjustment. It's a living ecosystem that, like any garden, needs your love and care to thrive. And trust me, the rewards are well worth the effort.

Let's not forget the practical benefits, either. Growing your own food in a greenhouse can lead to substantial savings on your grocery bills. Plus, you have the peace of mind knowing exactly how your food was grown, free from pesticides and harmful chemicals.

In an era where food security and sustainability are becoming increasingly important, having a greenhouse is a step towards self-sufficiency. It's an empowering feeling to know that you can provide for yourself and your family, no matter what challenges the world may throw your way.

As we journey together through the chapters of this book, you'll find that greenhouse gardening is not just about plants; it's about a lifestyle. It's about cultivating patience, nurturing curiosity, and embracing the joy of discovery. We'll explore the nuts and bolts of greenhouse construction, from choosing the right materials to understanding the best layout for your needs.

We'll delve into the specifics of soil preparation, seed selection, and plant care, breaking down each process into simple, manageable steps. And for those who are ready to take their skills to the next level, we'll cover advanced topics like hydroponics, pest control, and maximizing yield.

But beyond the technical aspects, I want to share the stories and experiences that make greenhouse gardening so special. Like the time I grew a sunflower in the middle of winter, just to see if I could. Or the joy of harvesting the first ripe strawberry of the season, a taste so sweet it made all the work worthwhile. These are the moments that make gardening not just a hobby, but a passion. And it's a passion I can't wait to share with you.

Overview of the Book's Structure

As you embark on your greenhouse gardening journey, understanding the structure of this book will help you navigate through the wealth of information provided. This book is designed to guide you through every stage of greenhouse gardening, from initial planning to advanced techniques. Let me walk you through the structure, so you know exactly what to expect.

Part 1: Introduction to Greenhouse Gardening lays the foundation for your greenhouse adventure. We start by exploring the benefits of greenhouse gardening, including how it can transform your gardening experience by providing year-round growing opportunities. We'll discuss the different types of greenhouses available and help you choose the one that best suits your needs. Finally, we'll give you an overview of what's to come in the book, setting the stage for a successful greenhouse gardening journey.

Part 2: Building Your Greenhouse covers the practical aspects of constructing your greenhouse. We begin with the essential tools and materials you'll need, ensuring you're well-equipped for the task. The following chapters provide detailed, step-by-step instructions for building a basic greenhouse, with additional techniques for more advanced constructions. Whether you're a DIY enthusiast or planning a more complex setup, this section will guide you through the process with clear, actionable advice.

Part 3: Creating the Perfect Greenhouse Environment focuses on optimizing the conditions within your greenhouse. We delve into climate control, including temperature, humidity, and light management. Proper ventilation and air circulation are crucial for plant health, and we'll cover the best practices for heating and cooling your greenhouse. This part ensures you can create an ideal growing environment for your plants, enhancing their growth and productivity.

Part 4: Growing Plants in Your Greenhouse provides practical guidance on selecting and growing plants in your greenhouse. We'll explore how to choose the right plants, start seeds, and propagate new plants. Detailed planting and growing techniques will help you care for your greenhouse plants, ensuring they thrive and yield abundant produce. This section also covers maintenance and troubleshooting, so you're prepared to handle any challenges that arise.

Part 5: Advanced Greenhouse Gardening Techniques introduces you to more sophisticated methods of greenhouse gardening. Hydroponics and aquaponics are explored as alternative growing systems, offering innovative ways to cultivate plants. We'll also discuss greenhouse automation, which can streamline your gardening tasks and improve efficiency. This part is for those looking to take their greenhouse gardening to the next level.

Part 6: Greenhouse Gardening Beyond the Basics expands on the foundational knowledge with advanced topics. Sustainable practices will help you reduce your environmental impact and make your greenhouse operation more eco-friendly. We'll also explore ways to expand your greenhouse operations, whether for additional crops or as a small-scale business. Finally, we'll cover harvesting and preserving your crops, ensuring you can enjoy the fruits of your labor long after the growing season ends.

By following the structure outlined in this book, you'll gain comprehensive knowledge and practical skills to make the most of your greenhouse gardening experience. Each section is designed to build upon the previous one, providing you with a thorough understanding of greenhouse gardening from start to finish.

As you read this book, I hope you'll find inspiration and encouragement. Whether you're looking to start small with a few potted herbs or planning to build a large, productive greenhouse, there's something here for everyone. Greenhouse gardening is a journey, one that offers endless opportunities for learning and growth. It's a path that leads not only to a bountiful harvest but also to a deeper connection with nature and yourself.

So, are you ready to get your hands dirty? Are you excited to embark on this green adventure? I promise it's going to be a journey full of wonder, learning, and, of course, delicious homegrown produce. Let's open the doors to your greenhouse and step into a world where the possibilities are as limitless as your imagination. Welcome to the world of greenhouse gardening—your personal paradise, your canvas for creativity, and your gateway to a healthier, more sustainable lifestyle.

As we dig deeper into the chapters ahead, remember that every great garden starts with a single seed. And every great gardener was once a beginner. Whether you're here to grow food and flowers or simply to find a new hobby, I'm thrilled to be your guide.

Let's make your greenhouse garden not just a reality but a flourishing success. Together, we'll create a space that's not just a garden but a testament to the beauty and resilience of nature and a reflection of your own journey towards growth and renewal. Welcome to the adventure of a lifetime. Let's get growing!

PART I: INTRODUCTION TO GREENHOUSE GARDENING

CHAPTER 1

Welcome to Year-Round Gardening!

"The garden of the future thrives not just in open fields but within the shelter of our greenhouses, where each season's promise endures."

Imagine a world where you can nurture your favorite plants in every season, regardless of the weather outside. That's the magic of greenhouse gardening. As you step into your greenhouse, you'll leave behind

the unpredictable whims of nature and enter a controlled environment where you dictate the terms of growth.

Whether it's the middle of winter or the peak of summer, your greenhouse becomes a sanctuary for your plants, offering them the perfect conditions to thrive year-round.

The Benefits of Greenhouse Gardening

Greenhouse gardening opens up a world of possibilities for gardeners by providing a controlled environment where plants can flourish. One of the primary benefits is the ability to enjoy year-round fresh produce. Imagine harvesting juicy tomatoes, crisp cucumbers, and fragrant herbs even in the depths of winter.

This is not just a luxury but a reality that greenhouse gardening makes possible. By shielding plants from harsh weather, greenhouses allow you to cultivate fresh, homegrown produce throughout the year. I've experienced firsthand how the taste of homegrown vegetables grown in a greenhouse far surpasses anything from the store. The flavor, the texture, and the freshness are unmatched.

Extended growing seasons are another significant advantage. In a greenhouse, you can start your seeds earlier in the spring and keep your plants growing later into the fall. This extended season allows for multiple harvests and the ability to grow crops that would otherwise be limited by your local climate. For instance, you can grow tender greens like lettuce and spinach long after the outdoor growing season has ended. I've found that this flexibility not only maximizes my yield but also lets me experiment with different plant varieties and growing techniques.

Protection from pests and extreme weather is a crucial benefit of greenhouse gardening. In a greenhouse, you create a barrier that keeps pests like aphids and caterpillars at bay. You also shield your plants from heavy rains, damaging winds, and freezing temperatures. This protection reduces the need for chemical pesticides and fertilizers, contributing to a healthier growing environment. Over the years, I've seen how this natural protection not only improves plant health but also simplifies garden maintenance.

Overall, greenhouse gardening enhances your gardening experience where you have greater control over growing conditions. This control translates into higher yields, better-quality produce, and a more satisfying gardening journey.

Types of Greenhouses

When it comes to greenhouse gardening, selecting the right type of greenhouse is crucial. Each design offers different advantages and is suited to various gardening needs. Let's explore the most common types and their unique benefits.

Attached Greenhouses are built onto the side of a building, usually a house. This design allows you to take advantage of the building's existing heating and cooling systems, which can significantly reduce the cost of maintaining the greenhouse.

Attached greenhouses are ideal for those who want to integrate their greenhouse into their home's overall design, providing easy access and convenience. I've found that having attached greenhouses allows better temperature regulation and makes it easier to monitor and care for plants.

Freestanding Greenhouses are standalone structures that can be placed anywhere on your property. They are flexible enough to be placed at a location where they'll receive the best sunlight and airflow.

Freestanding greenhouses are perfect if you have the space and want a dedicated area for gardening. They can be expanded or modified to expand more easily than attached greenhouses.

Lean-to Greenhouses are built against an existing wall, such as a garage or shed. This design maximizes space by using the wall for structural support and heat retention. Lean-to greenhouses are ideal for smaller properties or for gardeners who want to use the greenhouse as an extension of their existing garden space. They are energy-efficient, as they benefit from the wall's insulation.

Ridge-and-Furrow Greenhouses are designed with multiple sections connected together. This type is often used in commercial settings but can be adapted for larger home gardens. The ridge-and-furrow design allows for efficient use of space and energy, making it easier to manage multiple growing areas.

Dome Greenhouses have a unique, rounded shape that provides excellent light distribution and can withstand heavy snow loads. The dome shape helps reduce heating costs and offers a distinctive aesthetic in the garden.

Hoop Houses, or high tunnels, are a more affordable and simpler option. These structures are made with a framework of metal or plastic pipes covered with a single layer of plastic. Hoop houses are excellent for extending the growing season and protecting plants from frost. They are a great choice if you're just starting out and want a low-cost, effective solution for greenhouse gardening.

Each type of greenhouse has its advantages and can be tailored to fit different gardening needs and budgets. When choosing the right greenhouse for you, consider factors like your available space, budget, and the specific needs of your plants. I've had experience with several types of greenhouses, and each has its unique benefits. The key is to select the one that aligns with your gardening goals and provides the best environment for your plants.

Year-Round Fresh Produce

One of the most appealing aspects of greenhouse gardening is the ability to enjoy fresh produce year-round. The notion of harvesting ripe, homegrown vegetables and herbs in the middle of winter is not just a fantasy; it's an achievable goal with the right setup. Let me share with you how greenhouse gardening can provide you with a continuous supply of fresh produce, regardless of the season.

In a greenhouse, you control the environment, which means you can create ideal growing conditions for your plants throughout the year. The absence of harsh weather conditions like frost, snow, and heavy rains ensures that your plants are protected and can grow uninterrupted. This protection allows you to grow a wide variety of crops that might not thrive in your local outdoor climate.

I've personally experienced the joy of harvesting fresh greens like spinach and lettuce in the middle of winter. These leafy greens are incredibly versatile and can be used in a variety of dishes, from salads to soups. Growing them in a greenhouse means you don't have to rely on store-bought, often wilted greens that have traveled long distances. The freshness of homegrown greens is unmatched, and the taste is simply superior.

In addition to leafy greens, greenhouse gardening allows you to grow a range of other vegetables and herbs. Tomatoes, peppers, cucumbers, and herbs like basil and cilantro thrive in a greenhouse environment. With careful planning and management, you can even grow crops like strawberries and dwarf fruit trees. The key is to understand the specific requirements of each plant and ensure they receive adequate light, water, and nutrients.

Harvesting fresh produce year-round not only enhances your meals but also contributes to a healthier lifestyle. Homegrown produce is free from pesticides and other chemicals commonly found in commercially grown vegetables. You have full control over what goes into your plants, ensuring that you and your family enjoy clean, nutritious food.

You can experiment with different varieties and growing techniques by greenhouse gardening. You can try growing heirloom vegetables, rare herbs, or exotic fruits that might not be available in your local stores. This experimentation adds excitement to your gardening routine and expands your culinary horizons.

Extended Growing Seasons

One of the greatest advantages of greenhouse gardening is the ability to extend your growing seasons. In a traditional outdoor garden, the growing season is dictated by the weather and climate, with a clear start and end. However, in a greenhouse, you can effectively lengthen both the spring and fall seasons, allowing for a more productive and flexible gardening experience.

I'll explain how it works. By providing a controlled environment, greenhouses mitigate the impact of weather extremes. This control means you can start planting earlier in the spring and continue growing later into the fall. This extended growing window allows me to maximize my garden's productivity and experiment with crops that would otherwise be limited by seasonal constraints.

Starting plants earlier in the spring is particularly beneficial for crops that require a longer growing period. For instance, tomatoes and peppers need several months of warm weather to mature fully. By starting these plants in your greenhouse before the last frost date, you give them a head start, leading to an earlier and more abundant harvest. I've seen how this early start can make a significant difference in the size and quality of the produce.

Likewise, extending the growing season into the fall can provide a second harvest for many crops. Leafy greens, root vegetables, and brassicas like kale and Brussels sprouts thrive in cooler temperatures and can continue growing well into the fall. With a greenhouse, you can protect these plants from early frosts and enjoy fresh, homegrown produce long after winter has started outside.

Additionally, a greenhouse allows you to stagger your plantings. This technique involves planting seeds or seedlings at different times to ensure a continuous harvest. For example, you can sow successive batches of lettuce or radishes to enjoy a steady supply of fresh greens throughout the growing season. This approach also helps to minimize waste and maximize the use of available space.

By extending your growing seasons, you also have the opportunity to grow a wider variety of crops. Certain plants, like melons and pumpkins, require a long growing season to reach maturity. In a greenhouse, you can provide the warmth and light these plants need to thrive, even in regions with shorter outdoor growing seasons. You can expand your gardening repertoire using all these ways.

Protection from Pests and Weather

Greenhouses are a valuable tool which shield your plants from external threats, resulting in healthier produce and a satisfying gardening experience

In an outdoor garden, pests like aphids, caterpillars, and snails can cause significant damage to your crops. Greenhouses act as a barrier, preventing these pests from reaching your plants. This reduces the need for chemical pesticides, and I've witnessed that it simplifies my garden maintenance.

Greenhouses shield plants against extreme weather conditions such as heavy rains, strong winds, and freezing temperatures. They create a stable environment, where growth isn't hindered due to the harsh weather conditions outside. Delicate seedlings and tender plants can thrive in a greenhouse this way.

Greenhouses also help regulate temperature and humidity levels, further enhancing plant protection. This ideal growing environment minimizes stress on your plants which gives higher yields of better quality. I've observed that plants grown in a well-managed greenhouse are generally more robust and productive than those exposed to fluctuating outdoor conditions.

The controlled environment reduces the risk of soil erosion and allows for better management of soil health. You can focus on providing the right nutrients and growing conditions without worrying about external factors impacting your plants.

Overall, the protection provided by a greenhouse from pests and extreme weather enhances your gardening experience and contributes to healthier, more productive plants.

CHAPTER 2

Planning Your Greenhouse Dream

"To plant a garden is to believe in tomorrow." — Audrey Hepburn

Planning your greenhouse is where your gardening dream starts to take shape. This is the stage where you get to envision what your greenhouse will look like, how it will function, and how it will help you achieve your gardening goals.

As someone who has been gardening for many years, I can tell you that taking the time to plan thoroughly will save you countless hours of trouble down the road.

Key Considerations

When you're planning your greenhouse, there are several key considerations to keep in mind. These will serve as the foundation upon which your greenhouse will be built, ensuring that it meets your needs and supports your gardening ambitions.

First and foremost, think about the purpose of your greenhouse. Are you aiming to grow vegetables, flowers, or perhaps both? Different plants have different requirements, and understanding what you want to grow will influence many of your decisions. For instance, if you're focusing on vegetables, you'll need plenty of light and space for larger plants like tomatoes and cucumbers. If you're into flowers, you might need different temperature and humidity levels.

Next, consider the size of your greenhouse. This is often dictated by the space you have available and your budget. A larger greenhouse will give you more room to grow a variety of plants, but it will also require more resources to build and maintain. On the other hand, a smaller greenhouse can be more manageable and cost-effective, especially if you're just starting out.

Location is another critical factor. You want your greenhouse to be in a spot that gets plenty of sunlight, as most plants need at least six hours of direct sunlight a day. Additionally, consider the proximity to water sources and electricity if you plan to install heating, cooling, or automatic watering systems. Accessibility is important, too; you don't want to trek across your yard every time you need to tend to your plants.

Ventilation is crucial for maintaining a healthy greenhouse environment. Proper airflow helps regulate temperature and humidity, preventing the buildup of excess moisture that can lead to mold and mildew. Think about how you'll incorporate vents, fans, or other ventilation systems into your greenhouse design.

Lastly, think about future growth. You might start small, but as your gardening skills and ambitions grow, you may want to expand your greenhouse.

Choosing the Right Materials

Selecting the right materials for your greenhouse is a key step that will determine its durability, efficiency, and cost. There are several options available, each with its own set of advantages and disadvantages. Let's break down the most common materials used in greenhouse construction.

For your greenhouse structure, you can choose from wood, metal, or PVC. Wood is popular for its aesthetic appeal, ease of construction, and affordability, but it requires proper maintenance to prevent warping or deterioration. Metal, especially aluminum, is lightweight, strong, and rust-resistant, though it can be more expensive. PVC offers affordability and ease of assembly, with good durability and moisture resistance, but it may lack the sturdiness of wood or metal for larger structures.

When it comes to covering your greenhouse, you have three main options: glass, polycarbonate, and polyethylene film. Glass is the traditional choice, known for its clarity and longevity. It allows for maximum light transmission but can be heavy and expensive. It's also prone to breaking, which can be a concern in areas with high winds or hail.

Polycarbonate panels are a modern alternative to glass. They are lightweight, durable, and provide good insulation. They come in single, double, or triple layers, with more layers offering better insulation. The downside is that they can yellow over time and are generally more expensive than polyethylene film.

Polyethylene film is the most cost-effective option. It's easy to install and provides good light transmission. However, it's not as durable as glass or polycarbonate and may need to be replaced every few years. A double-layer polyethylene film with an air layer in between can improve insulation and extend its lifespan.

In addition to the structural and covering materials, consider the materials for the flooring and benches inside your greenhouse. Gravel, concrete, or pavers are good choices for flooring as they provide good drainage and can help regulate temperature. For benches, treated wood or metal are durable options that can support the weight of your plants and gardening tools.

Designing Your Greenhouse

Designing your greenhouse is where you get to put your personal touch on your gardening space. A well-designed greenhouse not only looks good but also functions efficiently, making your gardening tasks easier and more enjoyable.

Start with the layout. Think about how you'll use the space inside your greenhouse. You'll need paths wide enough to move comfortably, space for workbenches or potting tables, and room for your plants to grow. Consider incorporating shelves or vertical gardening systems to maximize space, especially if you have a smaller greenhouse.

Ventilation is an important aspect of greenhouse design. You'll need to include vents, windows, or fans to ensure proper airflow. Roof vents are particularly effective as they allow hot air to escape from the top of the greenhouse. Side vents or windows can also help by bringing in fresh air from the outside. If you live in a hot climate, you might also want to consider installing shade cloths or blinds to prevent your greenhouse from overheating.

Heating and cooling are other key considerations. Depending on your climate, you might need to install heaters, fans, or cooling systems such as air conditioning units to maintain a consistent temperature.

Lighting is crucial for plant growth, especially during the shorter days of winter. Natural light is ideal, so place your greenhouse where it will get the most sunlight. You can also supplement with artificial lights if needed. LED grow lights are energy-efficient and can provide the full spectrum of light that plants need.

Watering systems can make your life much easier. Drip irrigation systems are efficient and ensure that your plants get the right amount of water without wastage. You can also install automatic timers to water your plants at specific intervals, freeing up your time for other gardening tasks.

Budgeting for Your Greenhouse

Budgeting for your greenhouse is a critical step that requires careful planning and consideration. The costs can vary widely depending on the size, materials, and features you choose. By setting a realistic budget and sticking to it, you can avoid unexpected expenses and ensure your greenhouse project stays on track.

Start by estimating the cost of materials. This includes the structure, covering, flooring, benches, and any additional features like ventilation systems, heaters, and lighting. Get quotes from multiple suppliers to find the best prices, and don't forget to factor in the cost of delivery.

Next, consider the cost of tools and equipment. If you're building the greenhouse yourself, you'll need basic tools like hammers, saws, drills, and levels. You might also need specialized tools for tasks like cutting metal or installing glazing. If you don't already own these tools, consider borrowing or renting them to save money.

Labor costs are another important consideration. If you're hiring a contractor to build your greenhouse, get multiple quotes and check references to ensure you're getting a fair price. If you're doing the work yourself, factor in the value of your time. Building a greenhouse can be a significant time investment, so make sure you're prepared for the commitment.

Don't forget about ongoing costs. This includes utilities like water and electricity, as well as maintenance and repairs. Heating and cooling can be particularly expensive, so consider energy-efficient options like solar heaters or insulated panels to keep costs down.

Set aside a contingency fund for unexpected expenses. No matter how well you plan, there's always the possibility of unforeseen costs. Having a buffer in your budget can help you handle these surprises without derailing your project.

Selecting the Ideal Location

Choosing the right location for your greenhouse is crucial to its success. The ideal spot will provide the right amount of sunlight, protection from the elements, and easy access to water and electricity. Here are some key factors to consider when selecting your greenhouse location.

Sunlight is the most important factor. Most plants need at least six hours of direct sunlight a day, so choose a location that gets plenty of sun. In the northern hemisphere, a south-facing location is ideal, as it will receive the most sunlight throughout the year. Avoid areas with large trees or buildings that could cast shadows on your greenhouse.

Consider the proximity to water and electricity when planning your greenhouse. A reliable water source is essential for irrigation, and access to electricity is crucial if you're using heaters, fans, or grow lights. The closer your greenhouse is to these utilities, the easier and more cost-effective it will be to set up and maintain.

If you're in a remote area or want to be more sustainable, you might also consider installing solar panels. Solar panels can provide a renewable energy source, reducing long-term electricity costs and environmental impact. However, they can be expensive to install initially and may require consistent sunlight to be effective.

Protection from the wind is another important consideration. Strong winds can damage your greenhouse and make it difficult to maintain a stable temperature. Choose a location that's sheltered from prevailing winds, either by natural barriers like trees or hills or by building windbreaks.

The ground should be level and well-drained. Poor drainage can lead to waterlogging and root rot, while a sloped site can make construction more difficult. If your chosen site isn't level, you may need to do some grading or install a raised foundation to create a stable base for your greenhouse.

Think about accessibility. You'll need to access your greenhouse regularly to tend to your plants, so choose a location that's easy to reach. Consider the path you'll take to get there, especially during bad weather. A paved or gravel path can make it easier to transport tools and supplies to and from your greenhouse.

Customizing for Climate Needs

Every climate presents unique challenges for greenhouse gardening, and customizing your greenhouse to meet these needs is essential for success. Whether you're dealing with extreme temperatures, high humidity, or frequent storms, there are strategies and modifications you can implement to create a stable and productive growing environment.

To combat overheating in hot climates, consider installing shade cloths or reflective paint to reduce the amount of sunlight entering your greenhouse. Ventilation is also crucial; roof vents, side vents, and fans can help expel hot air and bring in cooler air. Evaporative coolers or misting systems can lower the temperature further by using the cooling effect of water evaporation.

In cold climates, insulation and heating are key. Double or triple-layered polycarbonate panels can provide good insulation while allowing plenty of light. You can also use thermal screens or insulated curtains to keep heat in during the night. For heating, consider options like electric heaters, gas heaters, or wood-burning stoves. Solar heaters can be an eco-friendly alternative, using the sun's energy to warm your greenhouse.

High humidity can lead to problems with mold, mildew, and pests. To manage humidity levels, ensure your greenhouse has adequate ventilation. Dehumidifiers can also help reduce moisture in the air. In some cases, you might need to limit watering to prevent excess humidity from building up.

Storm protection is another important consideration, especially in areas prone to strong winds, heavy rain, or hail. Choose a sturdy structure and secure it well to the ground. Reinforce the frame and cover with additional supports if necessary. If hail is a concern, opt for hail-resistant materials like polycarbonate panels instead of glass.

Finally, consider your local climate's seasonal variations. In areas with distinct seasons, you may need to adjust your greenhouse setup throughout the year. This could include adding or removing insulation, changing ventilation settings, or using different types of shade cloth. You can create a greenhouse environment that supports healthy plant growth year-round by being productive and adaptable.

PART 2: BUILDING YOUR GREENHOUSE

CHAPTER 3

Tools and Materials for Greenhouse Construction

"A well-built greenhouse starts with the right tools and materials, carefully chosen for quality and durability." — Anonymous.

Embarking on the journey of constructing your own greenhouse is both exciting and rewarding. With the right preparation, you can create a space that

will not only extend your growing season but also protect your plants from the elements.

Over the years, I've found that having the correct tools and materials can make the difference between a smooth building process and one filled with challenges.

Essential Tools

When it comes to building a greenhouse, having the right tools at your disposal can make the process much more efficient and enjoyable. While the specific tools you'll need can vary depending on the type and size of the greenhouse you're building, there are some essentials that every gardener should have in their arsenal.

First and foremost, a reliable set of measuring tools is crucial.

Tools and Materials you need include:

- **Tape Measure**
- **Long Carpenter's Level**
- **Framing Square**

The above are indispensable for ensuring your structure is accurate and level. Precise measurements will help you avoid costly mistakes and ensure that your greenhouse is sturdy and well-constructed.

You'll also need a good set of cutting tools. A handsaw or circular saw is essential for cutting wood. If you're working with metal, a hacksaw or angle grinder with a metal-cutting blade will come in handy. For cutting greenhouse coverings like polycarbonate panels or polyethylene film, a utility knife or tin snips are the tools of choice. Always make sure your cutting tools are sharp and in good condition to make clean, precise cuts.

Fastening tools are another key component of your toolkit. A good drill with a variety of bits is essential for drilling holes and driving screws. If you're working with metal, a rivet gun might be necessary for securing metal panels. A set of screwdrivers, both flathead and Phillips, will also be useful for assembling parts and making adjustments.

A set of wrenches and pliers is important for tightening bolts and nuts, as well as bending and holding materials in place. Adjustable wrenches are versatile and can be used on a variety of fastener sizes, while needle-nose pliers are great for reaching into tight spaces.

For moving and handling materials, a wheelbarrow, dolly, or hand truck can save you a lot of time and effort. These tools are particularly useful for transporting heavy items like bags of concrete, rolls of film, or large panels.

Safety gear is also essential. Always wear a pair of sturdy work gloves to protect your hands from cuts and splinters. Safety glasses or goggles will protect your eyes from debris, and a dust mask or respirator is important when cutting materials that produce dust, such as wood or polycarbonate.

Don't underestimate the value of a good workbench or table. Having a stable surface to work on can make tasks like measuring, cutting, and assembling much easier and safer. Make sure your workbench is sturdy and at a comfortable height to minimize strain on your back and shoulders.

Selecting Building Materials

Choosing the right materials for your greenhouse construction is a critical step that will influence the durability, efficiency, and aesthetics of your final structure. With so many options available, it can be overwhelming to decide what materials to use. However, by considering factors such as climate, budget, and the types of plants you want to grow, you can make informed choices that will set you up for success.

The frame of your greenhouse is the backbone of the structure, and it needs to be strong and durable. Common materials for greenhouse frames include wood, metal, and PVC. Wood is a popular choice for its natural look and ease of construction. It's important to use treated wood or naturally rot-resistant species like cedar or redwood to ensure longevity. However, wood can be susceptible to warping and decay if not properly maintained.

Metal frames, typically made from aluminum or galvanized steel, offer strength and durability. Aluminum is lightweight and resistant to rust, making it a great choice for greenhouse construction. Galvanized steel is even stronger and can support larger structures, but it's heavier and may require more specialized tools for assembly. Both materials are low-maintenance and can withstand harsh weather conditions.

PVC is another option known for its affordability and ease of use. It's lightweight, resistant to moisture, and easy to assemble with basic tools. However, PVC may not be as strong as wood or metal and can become brittle over time, especially in cold climates. It's best suited for smaller greenhouses or temporary structures.

The covering material is another crucial component of your greenhouse. It determines how much light and heat enters the structure and how well it retains warmth. The most common options are glass, polycarbonate panels, and polyethylene film.

Glass is the traditional choice and offers excellent light transmission and durability. However, it's heavy, expensive, and can be prone to breaking. If you choose glass, consider using tempered or double-pane options for added strength and insulation.

Polycarbonate panels are a popular alternative to glass. They're lightweight, durable, and provide good insulation. They come in single, double, or triple layers, with more layers offering better insulation. Polycarbonate panels are also shatter-resistant, making them a safer option in areas prone to hail or high winds.

Polyethylene film is the most cost-effective option. It's easy to install and provides good light transmission. Double-layered polyethylene with an air space between layers offers better insulation than single-layer film. However, polyethylene is less durable than glass or polycarbonate and may need to be replaced every few years. UV-resistant options are available to extend the lifespan of the film.

For the foundation, you can use materials like concrete, gravel, or treated wood. A concrete foundation is the most durable and provides a solid base for your greenhouse. It's especially useful in areas with strong winds or heavy snowfall. Gravel is a more affordable option that provides good drainage and is easy to install. Treated wood can also be used for a simple, raised foundation.

Safety Precautions

Safety should always be a top priority when constructing a greenhouse. While building your own greenhouse can be a fulfilling project, it also involves working with tools, heavy materials, and potentially hazardous conditions. By taking the necessary precautions, you can minimize risks and ensure a safe construction process.

Safety Tools you need:

- Safety Glasses or Goggles
- Protective Gloves
- Ear Protection (e.g., Earplugs or Earmuffs)

Always use tools according to the manufacturer's instructions and ensure they are in good working condition. Check for frayed cords, dull blades, or loose parts, and replace or repair any damaged equipment. Keep your tools clean and well-maintained to reduce the risk of accidents.

When lifting heavy materials, use proper lifting techniques to avoid back injuries. Bend at the knees, keep your back straight, and lift with your legs, not your back. If a material is too heavy to lift safely on your own, don't hesitate to ask for help or use equipment like a dolly or wheelbarrow.

Be mindful of your surroundings and keep your workspace organized. Clear any obstacles that could cause trips or falls, and ensure that tools and materials are stored safely when not in use. If you're working at heights, such as on a ladder, make sure the ladder is stable and placed on a level surface. Avoid overreaching, and always have someone nearby to assist if needed.

Electrical safety is another important consideration, especially if you're installing lighting, fans, or heaters in your greenhouse. Ensure all electrical work is done according to local codes and regulations. Use weatherproof outlets and enclosures, and keep electrical cords away from water sources to prevent electrical shock.

When working with chemicals, such as paints, sealants, or pesticides, always read the labels and follow safety instructions. Use these substances in a well-ventilated area, and wear protective clothing and a mask if necessary. Store chemicals safely out of reach of children and pets, and dispose of any waste materials according to local regulations.

Be prepared for emergencies. Keep a first aid kit at hand, and know basic first aid procedures. Have a plan in place for dealing with accidents or injuries, and know the location of the nearest medical facility. If you're working with others, make sure everyone is aware of the safety procedures and emergency plans.

Sourcing Materials

Finding the right materials for your greenhouse construction is an important part of the planning process. The quality and availability of materials can vary greatly, so it's important to know where to look and what to consider when making your selections. Here are some tips for sourcing the best materials for your project.

Local hardware stores and home improvement centers are a great starting point for finding basic construction materials like wood, metal, and PVC. In the United States, Home Depot and Lowe's are two of the largest chains, offering a wide range of products and helpful advice. In the United Kingdom, B&Q and Wickes are popular choices, while in Canada, Home Hardware and RONA are well-known options. These stores often carry greenhouse-specific materials like polycarbonate panels and polyethylene film, making them convenient for both beginners and experienced builders.

Specialty suppliers and online retailers are excellent resources for more specialized materials. These suppliers often offer a wider selection of greenhouse-specific products, including custom-cut panels, high-quality coverings, and specialized ventilation or heating equipment. Websites like Amazon, eBay, or dedicated gardening and greenhouse supply sites can provide access to a broader range of materials and customer reviews to help you make informed decisions.

If you're looking for more affordable options, consider checking out salvage yards, thrift stores, or online marketplaces like Craigslist or Facebook Marketplace for used materials. You can often find reclaimed wood, metal, or glass at a fraction of the cost of new materials. Just be sure to inspect these items carefully for any damage or defects that could compromise the safety or durability of your greenhouse.

Local building supply auctions or clearance sales can also be a good source of discounted materials. These events often feature surplus or discontinued items at reduced prices. Keep an eye out for sales at local stores or check online for auction listings in your area.

When selecting materials, quality should always be a top consideration. While it may be tempting to choose the cheapest option, investing in high-quality materials will pay off in the long run. Quality materials are more durable, easier to work with, and less likely to require repairs or replacements.

Another important factor to consider is the environmental impact of your materials. Whenever possible, choose sustainable and eco-friendly options. For example, opt for FSC-certified wood, which comes from responsibly managed forests, or recycled materials. Additionally, consider the energy efficiency of your materials, such as using double-layered polycarbonate panels for better insulation.

When sourcing materials, it's also important to consider the size and specifications of your greenhouse. Measure your space carefully and make sure you purchase the correct amounts of materials. Ordering a little extra can be a good idea to account for any mistakes or changes in your plans.

Don't forget about delivery and transportation. Some materials, like large panels or heavy bags of concrete, may require special handling or equipment to transport. Check with your supplier about delivery options and costs, and make sure you have a plan for moving and storing your materials once they arrive.

Building a Tool Kit

Having a well-equipped toolkit is essential for any DIY greenhouse project. A comprehensive toolkit not only makes the construction process smoother but also ensures that you have everything you need to tackle any challenges that arise. In this section, we'll discuss the essential tools to include in your toolkit and how to organize them for maximum efficiency.

The foundation of any good toolkit is a reliable set of measuring tools. A tape measure, at least 25 feet long, is essential for measuring materials and ensuring accuracy in your cuts. A carpenter's level is crucial for making sure your structure is straight and level, which is especially important for the stability and longevity of your greenhouse. A framing square is useful for checking corners and ensuring they are perfectly square.

For cutting materials, you'll need a variety of saws. A handsaw is great for cutting wood, while a circular saw can make the job faster and easier, especially for larger pieces. If you're working with metal, a hacksaw or an angle grinder with a metal-cutting blade is necessary. For cutting greenhouse coverings, a utility knife with a sharp blade will provide clean, precise cuts.

Drilling and fastening tools are another critical component of your toolkit. A cordless drill with a set of drill bits is versatile and convenient for drilling holes and driving screws. Make sure to have both Phillips and flathead screwdriver bits, as well as a set of standard screwdrivers for manual adjustments. If you're working with metal, a rivet gun can be useful for securing panels. A set of pliers, including needle-nose and slip-joint pliers, will come in handy for gripping and bending materials.

Wrenches are essential for assembling and tightening bolts and nuts. An adjustable wrench is versatile and can be used on a variety of fastener sizes. A socket wrench set is also useful for working with bolts and nuts, providing more torque and making the job easier. Don't forget to include a set of Allen wrenches (hex keys) for any fasteners that require them.

For marking and measuring, include a pencil or marker, a carpenter's pencil, and a chalk line. These tools are useful for marking measurements and guidelines on your materials, ensuring precision in your cuts and assembly.

Other useful tools to have on hand include a mallet or hammer for driving nails or tapping materials into place, a staple gun for securing plastic coverings, and clamps for holding materials together while you work. A pry bar or crowbar can be useful for removing old materials or making adjustments.

In addition to the tools themselves, consider the accessories and consumables you'll need, such as screws, nails, washers, and bolts. Having a variety of sizes and types on hand will save you time and prevent delays. A tool belt or tool bag is also a good investment, allowing you to keep your most-used tools within easy reach while you work.

Organizing your tools is key to maintaining an efficient workspace. Store your tools in a toolbox or tool chest with compartments for different types of tools. This not only keeps your tools organized but also protects them from damage. Labeling compartments or using a pegboard with hooks can make it easy to find the tool you need quickly.

Regular maintenance of your tools is essential to keep them in good working condition. Clean your tools after each use, especially cutting tools, to remove debris and prevent rust. Sharpen blades as needed and replace any worn or damaged parts. Keeping your tools in good shape will ensure they last longer and work more efficiently.

Preparing Your Workspace

Creating a dedicated workspace for your greenhouse construction project is crucial for both efficiency and safety. A well-organized workspace not only makes the construction process smoother but also helps prevent accidents and ensures that you have everything you need at hand. In this section, we'll discuss the key elements of preparing your workspace, from setting up your area to organizing your tools and materials.

The first step in preparing your workspace is choosing an appropriate location. Ideally, your workspace should be close to your greenhouse site to minimize the distance you need to transport materials and tools. Look for a flat, stable surface with enough space to accommodate your materials, tools, and any workbenches or tables you'll need. If possible, choose a shaded area to protect yourself and your materials.

Once you've chosen a location, it's time to set up your workspace. Start by clearing the area of any debris, obstacles, or tripping hazards. If you're working on grass or dirt, consider laying down a tarp or plywood sheets to create a stable surface. This will help prevent tools and materials from sinking into the ground and make it easier to move around.

Next, set up your workbenches or tables. A sturdy workbench is essential for tasks like measuring, cutting, and assembling materials. Make sure your workbench is at a comfortable height to reduce strain on your back and shoulders. If you don't have a workbench, a sturdy table or a pair of sawhorses with a plywood top can work as a temporary solution.

Organizing your tools and materials is key to maintaining an efficient workspace. Keep your tools within easy reach, either on your workbench, in a toolbox, or hung on a pegboard. Group similar tools together, such as cutting tools, fastening tools, and measuring tools, to make it easy to find what you need. Use bins, buckets, or shelves to store smaller items like screws, nails, and fasteners.

For larger materials like wood, metal, or greenhouse panels, designate a storage area where they can be easily accessed. Stack materials neatly and securely to prevent them from tipping over or causing injury. If space is limited, consider storing materials vertically to save space. Just make sure they are well-supported and won't fall over.

Lighting is another important consideration, especially if you're working in a garage or workshop. Good lighting is essential for safety and precision. Make sure your workspace is well-lit, with overhead lights or task lights focused on your work area. If you're working outdoors, take advantage of natural light during the day and consider setting up portable work lights for evening or overcast conditions.

Safety should always be a priority in your workspace. Make sure you have easy access to the safety gear. If you're working with power tools or electrical equipment, have a fire extinguisher nearby and know how to use it. Keep your workspace clean and free of clutter to reduce the risk of accidents.

Finally, consider the logistics of your workspace. Think about the flow of your work and arrange your tools and materials accordingly. For example, if you'll be cutting materials, set up your saw near your material storage area and have a clear path to your assembly area. This will help you work more efficiently and reduce unnecessary movement.

By taking the time to prepare your workspace properly, you'll create a safe, organized, and efficient environment for building your greenhouse. This preparation will not only make the construction process smoother but also ensure that you have a positive and enjoyable experience. With everything in place, you'll be ready to tackle your greenhouse project with confidence and ease.

CHAPTER 4
Building a Basic Greenhouse

"A garden without a greenhouse is like a chef without a kitchen." — Anonymous

Building your own greenhouse is one of the most rewarding projects a gardener can undertake. It's where your gardening journey can truly blossom, allowing you to extend your growing season, experiment with new plant varieties, and protect your crops from adverse weather. Over the years, I've had the pleasure of constructing several greenhouses, each time refining the process and learning new tricks.

Step by Step Process of Building a Simple Greenhouse

Building a greenhouse involves several key stages, each requiring careful planning and execution. The pre requisites of building a greenhouse and factors to be considered have been explained in chapter 3. In this chapter we'll discuss the steps of creating a greenhouse structure.

Constructing a simple greenhouse might seem daunting, but with the right guidance and patience, it's a manageable and enjoyable project. Here is how you can construct your own simple greenhouse. Here is a step-by-step breakdown of each stage.

1. **Building the Foundation of the Greenhouse Structure**

Start by choosing a suitable location. You want an area that gets plenty of sunlight throughout the day, is well-drained, and has easy access to water and electricity if needed. Also, consider the wind patterns in your area; placing your greenhouse in a sheltered spot can protect it from strong gusts that could damage the structure or disturb the plants inside.

Once you've selected the location, it's time to prepare the ground. Clear the area of any debris, rocks, or plants that could interfere with the construction. It's essential to work on a level surface, so take the time to level the ground properly. This will not only make construction easier but also ensure the stability and longevity of your greenhouse.

With the ground prepared, begin marking out the footprint of your greenhouse. Use stakes and string to outline the dimensions of the structure. This will serve as a guide as you build and help you ensure that everything is square and aligned. Double-check your measurements before moving on to avoid any costly mistakes later.

Next, build the foundation. It can be simple or complex depending upon your budget and needs. For a simple greenhouse, a gravel or concrete base works well. Gravel is easy to lay and provides excellent drainage, while concrete offers a more permanent and stable foundation. For a gravel foundation, excavate the area to a depth of about 4-6 inches, level the ground, add a layer of crushed stone or gravel, and compact it to create a firm base. For a more permanent solution, construct a concrete slab foundation by building a wooden frame, pouring the concrete, leveling the surface, and allowing it to cure completely.

If you choose gravel, create a border using treated wood or bricks to keep the gravel in place. For concrete, you'll need to build a frame to pour the concrete into. Make sure the foundation is level and well-compacted, as this will support the entire structure.

2. Assembling the Frames of the Greenhouse

Once the foundation is set, move on to constructing the frame. The frame is the skeleton of your greenhouse and needs to be strong enough to support the walls, roof, and any equipment you plan to install inside. You can use wood, metal, or PVC for the frame, depending on your budget, climate, and personal preference. Wood is easy to work with and provides a natural look, while metal is durable and can support heavier loads. PVC is lightweight and inexpensive but may not be as sturdy as other materials.

3. Assembling the Base Frame

Assemble the frame according to your design, making sure all the joints are secure and the structure is level. Use a carpenter's level to check your work frequently. If you're using wood, treat it with a weather-resistant finish to protect it from the elements. Metal frames should be galvanized or otherwise treated to prevent rust. If you're using PVC, make sure it's UV-resistant to avoid degradation over time.

The first stage is assembling the base frame, which forms the bottom perimeter of your greenhouse. This base frame provides the foundation for the rest of the structure and needs to be perfectly level and square. Start by laying out the frame pieces according to your plans, then secure them together with screws or bolts. If you're using wood, pre-drill the holes to prevent splitting. For metal or PVC, use appropriate connectors and fittings to join the pieces.

4. Installing the Vertical Supports

Next, you'll install the vertical supports, which form the sides of the greenhouse. These supports need to be strong enough to hold up the walls and roof. Attach them to the base frame, ensuring they are plumb (perfectly vertical) and evenly spaced. Use a level to check each support as you go. If you're working with wood, you can use metal brackets or angled braces to reinforce the joints. For metal or PVC frames, follow the manufacturer's instructions for securing the supports.

5. Adding the Horizontal Crosspieces

Once the vertical supports are in place, you can add the horizontal crosspieces, which will connect the supports and provide additional stability. These crosspieces are also where you'll attach the covering material later. Start with the lower crosspieces, then move up to the top, securing each piece with screws or bolts. Again, make sure everything is level and square. It's important to check your measurements frequently during this stage to ensure the structure is properly aligned.

6. Installing the Walls

Begin the installation of the walls by measuring and cutting the chosen covering material (glass, polycarbonate, or polyethylene). Install the wall panels starting from one corner, securing them to the frame with glazing clips, screws, or specialized mounting hardware. Ensure proper alignment and sealing at the joints to prevent air and water leaks.

7. Installing the Roof Supports

With the basic frame assembled, it's time to add the roof supports. The type of roof you choose will depend on your design and the materials you're using. A simple gable roof is easy to build and provides good headroom inside the greenhouse. For this type of roof, you'll need to install rafters that slope down from a central ridge beam to the side walls. Attach the ridge beam first, making sure it's level and centered, then secure the rafters to the ridge beam and the top of the vertical supports. Use metal connectors or brackets to reinforce the joints.

Install the roof panels by following the contours of the frame, especially for pitched roofs, starting from the bottom edge and working upwards. Overlap the panels to ensure the water runs off properly. Ensure a tight seal around the ridge and eaves, and consider incorporating ventilation features like vents or windows to promote effective airflow.

8. Customizing the Roof Style

If you're using a different roof style, such as a curved or flat roof, the process will vary slightly. For a curved roof, you'll need flexible framing materials, such as PVC or metal tubing, that can be bent into shape. For a flat roof, ensure that the roof is properly sloped to allow for water runoff. In all cases, the roof structure must be strong enough to support the weight of the covering material and any snow or wind loads in your area.

9. Installing the Door and Ventilation Windows

After the frame and roof are assembled, the next step is to install the door and any ventilation windows. The door should be sturdy and easy to open and close. Place it in a convenient location, ensuring it opens outward to maximize interior space. If you're installing windows, position them where they can provide adequate ventilation without compromising the structural integrity of the walls.

10. Adding Utilities and Sustainability

Before finalizing the build, consider the proximity to water and electricity. You'll need a reliable water source for irrigation and access to electricity if you're using heaters, fans, or grow lights. The closer your greenhouse is to these utilities, the easier and more cost-effective it will be to set up and maintain.

Additionally, if you live in a remote area or want to be more sustainable, consider installing solar panels to power your greenhouse. Solar panels can reduce energy costs and provide a renewable energy source, though they can be more expensive to install and may require maintenance over time.

Troubleshooting Tips

Even with careful planning and execution, building a greenhouse can sometimes present unexpected challenges. In this section, I'll share some common issues you might encounter during construction and offer practical solutions to help you troubleshoot and resolve them.

One common issue is dealing with an uneven foundation. If your foundation isn't perfectly level, it can cause the entire structure to be misaligned. To fix this, you may need to adjust the ground level or add shims under the base frame to level it out. Use a long carpenter's level to check the levelness across the entire foundation and make adjustments as needed. It's crucial to address this issue early, as a stable foundation is essential for the integrity of the entire greenhouse.

Another challenge can arise if your frame isn't square. An out-of-square frame can lead to problems with installing the walls and roof, as the panels won't fit properly. To check for squareness, measure the diagonals of the frame; they should be equal. If they're not, you'll need to adjust the position of the supports until the measurements are equal. It's a good idea to re-check the squareness at multiple stages of the build, especially before securing the crosspieces and roof supports.

You might also encounter issues with fitting the covering material. For example, if you're using polycarbonate panels, they need to fit snugly in the frame without gaps. If the panels are too large, you can trim them down using a fine-toothed saw or a utility knife. If they're too small, you may need to adjust the frame slightly or use a sealing material like silicone to fill in any gaps. Always handle covering materials carefully to avoid scratches or damage, especially with transparent materials like glass or polycarbonate.

Sometimes, you might find that the frame feels wobbly or unstable, especially if you're using lightweight materials like PVC. To improve stability, add additional bracing, such as diagonal braces in the corners or along the sides. These braces help distribute weight and prevent the frame from flexing under stress. For wooden frames, use metal brackets or braces to reinforce joints and connections.

Weather can also pose challenges during construction. If you're building in a windy area, it's essential to secure any loose materials and take extra precautions to anchor the frame. You can use ground anchors or concrete footings to secure the base frame. During the construction process, avoid working with large panels on windy days, as they can catch the wind and become difficult to handle safely.

Finally, if you encounter issues with the door or windows not fitting properly, check the frame alignment and make any necessary adjustments. Ensure that the door frame is square and that the door hardware is installed correctly. For windows, check that they open and close smoothly and that the seals are tight to prevent drafts and water leaks.

Final Touches and Inspections

With the main structure of your greenhouse complete, it's time to focus on the final touches and inspections to ensure everything is in top shape. These last steps are crucial for maximizing the efficiency and longevity of your greenhouse, as well as ensuring the safety and comfort of your plants.

Start by installing any additional features or equipment you plan to use inside the greenhouse. This might include shelving, benches, irrigation systems, or heating and cooling units. Consider the layout of your greenhouse and how you will use the space.

Install shelving or benches at a comfortable working height, ensuring they are sturdy and level. For irrigation, choose a system that suits your plants' needs, whether it's a simple drip system or a more complex misting setup. If you live in an area with extreme temperatures, consider adding a heater or fan to regulate the climate inside the greenhouse.

Next, inspect the entire structure for any potential issues or areas that need improvement. Check the frame, walls, and roof for stability and security. Ensure that all joints and connections are tight and that the covering material is securely fastened with no loose or damaged sections. Pay special attention to the door and any vents or windows, making sure they open and close smoothly and seal properly when closed. If you find any issues, address them promptly to prevent further problems.

It's also a good idea to test the ventilation and heating systems to make sure they are working correctly. Open and close the vents or windows to see how well they operate and monitor the temperature inside the greenhouse. If you have installed a heating or cooling system, test it to ensure it maintains the desired temperature range. Proper ventilation and temperature control are key to preventing issues like overheating, humidity buildup, or mold growth.

Another important step is to check the drainage around the greenhouse. Ensure that the ground slopes away from the structure to prevent water from pooling around the foundation. If necessary, install gutters or downspouts to direct rainwater away from the greenhouse. Proper drainage is essential for maintaining a stable environment inside the greenhouse and preventing water damage to the structure.

Finally, consider the aesthetic and functional aspects of your greenhouse. Add any finishing touches, such as trim, paint, or decorative elements, to enhance the appearance and blend the greenhouse into your garden. Install pathways or stepping stones inside and around the greenhouse to create a clean, organized space for walking and working. These finishing touches not only improve the look of your greenhouse but also make it a more enjoyable and practical space to work in.

Before you start using your greenhouse, take a final walk-through inspection. Make sure all systems are functioning properly, all tools and materials are cleaned up and organized, and the greenhouse is ready to support your gardening efforts. Take note of any maintenance tasks that may be needed in the future, such as checking the seals, cleaning the covering material, or servicing the ventilation system.

Building your own greenhouse is a significant achievement and a valuable addition to your gardening repertoire. With the structure complete and the final touches in place, you can now look forward to enjoying the benefits of year-round gardening, experimenting with new plants, and creating a thriving, productive space for your horticultural passions. Congratulations on your hard work and dedication to this project—you're now ready to start growing in your new greenhouse!

CHAPTER 5

Advanced Greenhouse Construction Techniques

"A greenhouse isn't just a structure; it's a space where imagination and innovation flourish alongside your plants."

When you reach a certain level of comfort and experience with greenhouse gardening, the basic structure can start to feel limiting. Perhaps you've already built your first greenhouse and now want to push the boundaries of what's possible. Or maybe you're dreaming big right from the start, eager to incorporate cutting-edge technology and unique design features.

Advanced greenhouse construction techniques offer a wealth of possibilities for those looking to elevate their gardening experience.

Complex Greenhouse Designs

When it comes to greenhouse construction, complexity often translates to functionality and beauty. While simple greenhouses serve their purpose, advanced designs can provide a more tailored environment for your plants and create a visually stunning addition to your garden. Let's dive into the world of complex greenhouse designs, where functionality meets aesthetics, and explore how you can bring your vision to life.

Complex greenhouse designs often include multi-level structures, multi-zone environments, and integrated systems for heating, cooling, and irrigation. These features allow you to create different climates within the same greenhouse, catering to a diverse range of plants with varying needs.

For instance, you might want a tropical zone for exotic plants and a temperate zone for your vegetable garden. Multi-level designs can also help you maximize space, using vertical gardening techniques and tiered shelving to grow more plants in the same footprint.

When planning a complex design, it's crucial to start with a clear vision of what you want to achieve. Consider the types of plants you want to grow and their specific requirements.

Sketch out your ideas, and don't hesitate to consult with a professional if you're unsure about the technical aspects. Remember, a well-designed greenhouse is an investment that can enhance your gardening experience and increase the value of your property.

In terms of materials, complex designs often require more robust and durable options. Choice of covering and framing materials impact the appearance as well as the functionality and maintenance needs of your greenhouse.

Incorporating Advanced Features

Once you have a design in mind, it's time to think about the advanced features you want to incorporate. These can range from automated systems to aesthetic enhancements, all aimed at making your greenhouse more efficient, comfortable, and visually appealing.

One popular feature is an automated climate control system. This system can include temperature sensors, automated vents, fans, and heaters, all working together to maintain the perfect growing conditions for your plants. With automation, you can set your desired temperature and humidity levels, and the system will adjust as needed. This not only saves you time and effort but also ensures that your plants are always in the best possible environment.

Another advanced feature is an integrated irrigation system. From simple drip lines to more sophisticated hydroponic setups, these systems can deliver water and nutrients directly to your plants, reducing water waste and ensuring consistent growth. You can even set up a rainwater collection system to make your greenhouse more sustainable. Additionally, incorporating LED grow lights can extend your growing season.

Don't forget about aesthetic enhancements. Incorporating features like decorative trim, stained glass, or custom paint can transform your greenhouse into a garden centerpiece.

Pre-Fabricated Greenhouses

For sourcing pre-fabricated greenhouses, here are some examples of companies and types of suppliers you might consider:

- **The Home Depot:** Offers a variety of pre-fabricated greenhouse kits from brands like Palram and Rion. These kits include everything needed for assembly and are available in different sizes and materials.
- **Lowe's:** Carries pre-fabricated greenhouse kits from brands such as Gardeners Supply Company and Outsunny. They provide options in various styles and sizes to fit different gardening needs.

- **Greenhouse Megastore:** Specializes in a wide range of greenhouse kits, including both small hobby greenhouses and larger commercial models. They offer various materials and features, from basic structures to more advanced systems.
- **Growers Supply:** Provides pre-fabricated greenhouses designed for both hobbyists and commercial growers. Their kits include options with different types of coverings and frame materials.
- **Amazon:** Lists pre-fabricated greenhouse kits from various manufacturers, allowing you to compare different models and read customer reviews. Brands available might include Quictent and ShelterLogic.

When selecting a pre-fabricated greenhouse, it's beneficial to check the product reviews, warranty details, and available customer support from these suppliers to ensure you choose a reliable and suitable option for your needs.

Integrating Technology

In the modern age, technology has revolutionized greenhouse gardening, making it easier and more efficient than ever before. Integrating technology into your greenhouse can help you monitor and control the environment, improve plant health, and reduce maintenance tasks. Let's explore some of the tech options you can incorporate into your advanced greenhouse setup.

Smart greenhouse systems are at the forefront of this technological revolution. These systems use sensors and controllers to monitor temperature, humidity, soil moisture, and even CO2 levels. The data collected is then used to automate various aspects of the greenhouse, such as ventilation, heating, cooling, and irrigation. Many of these systems can be controlled remotely via a smartphone app, giving you the flexibility to manage your greenhouse from anywhere in the world.

Another exciting technological advancement I think impacts plant growth a lot is the use of LED grow lights. Unlike traditional lighting, LED lights can be customized to emit specific wavelengths of light, optimizing photosynthesis and promoting plant growth. They are energy-efficient and produce less heat, reducing the need for additional cooling systems. Some advanced LED systems even offer programmable lighting schedules, mimicking natural sunlight patterns to enhance plant development. They are effective when the days are short too, as they can provide extended light to plants.

For those interested in hydroponics or aquaponics, technology can play a crucial role in maintaining the nutrient balance and water quality. Automated dosing systems can ensure that your plants receive the right nutrients at the right time, while water quality monitors can alert you to any issues that need attention. These systems can significantly reduce the manual labor involved in maintaining a healthy growing environment.

Expanding Your Structure

As your gardening ambitions grow, so might your need for more space. Expanding an existing greenhouse can be a cost-effective way to increase your growing area without building an entirely new structure. Whether you're looking to add more growing space, create separate zones for different plant types, or incorporate additional features, expanding your greenhouse can be a rewarding project.

The first step in expanding your greenhouse is to assess your current structure and decide what you want to add. This could involve extending the length or width of the greenhouse, adding a second story, or building adjacent structures like cold frames or shade houses. When planning an expansion, it's essential to consider the load-bearing capacity of the existing frame and foundation. You may need to reinforce these elements to support the additional weight.

Expanding your greenhouse also offers an opportunity to upgrade existing systems. For example, you might install a more advanced ventilation system or upgrade your irrigation setup to accommodate the increased number of plants. It's also a good time to revisit your lighting and climate control systems to ensure they are adequate for the expanded space.

Custom Greenhouse Projects

Finally, let's explore the exciting world of custom greenhouse projects. For the creative gardener, building a custom greenhouse offers the ultimate expression of individuality and innovation. Whether you're designing a greenhouse that doubles as an art studio, a space for hosting garden parties, or a specialized environment for rare and exotic plants, the possibilities are endless.

Custom greenhouse projects often start with a unique vision. Maybe you want a greenhouse that seamlessly blends with your home's architecture, or perhaps you're interested in sustainable construction methods using reclaimed materials. Whatever your dream, a custom project allows you to modify every aspect of the design to meet your specific needs and preferences.

Working with an architect or specialized greenhouse builder can help bring your vision to life. These professionals can provide valuable insights into design, materials, and construction techniques, ensuring that your custom greenhouse is both beautiful and functional. They can also help navigate any building codes or regulations that may apply to your project.

In addition to the aesthetic and functional benefits, custom greenhouses can also be a sound investment. A well-designed and thoughtfully constructed greenhouse can add value to your property, providing a unique selling point that sets your home apart.

As you embark on your custom greenhouse journey, remember that this is your opportunity to create a space that reflects your passion for gardening and your unique personality. Whether you're building a modest retreat or a grand conservatory, your custom greenhouse can be a place of beauty, creativity, and growth.

PART 3: CREATING THE PERFECT GREENHOUSE ENVIRONMENT

CHAPTER 6
Understanding Greenhouse Climate

"Creating the perfect climate for your greenhouse plants is like crafting a recipe—every ingredient must be just right for the best results."

Maintaining the ideal climate within your greenhouse is essential for the health and productivity of your plants. Unlike outdoor gardens, where you must work with the natural weather conditions, a greenhouse allows you to control the environment to a certain extent.

Temperature Control

I've experienced that investing in quality climate control tools and systems not only enhances plant health but also simplifies greenhouse management. Here's an overview of the essential tools I suggest for managing temperature, humidity, and light in your greenhouse.

- **Heating Systems**
- **Ventilation Systems**
- **Shading Systems**
- **Humidity Control**
- **Lighting Systems**
- **Climate Control Computers**
- **Monitoring Instruments**

We'll discuss these different types of tools according to the factors that need to be adjusted for a better greenhouse.

Thermometers, hygrometers, and light meters are fundamental tools for monitoring climate conditions. Digital versions of these instruments provide accurate readings and can be connected to climate control systems.

Investing in these tools and integrating them into your greenhouse management practices will help you create a stable and productive environment for your plants. Regular maintenance and adjustments based on real-time data will ensure that your greenhouse supports optimal plant growth and health.

Methods of Temperature Control

Temperature control is at the heart of successful greenhouse management. Plants, like people, have their preferred temperatures, and maintaining these conditions is crucial for their health and productivity. Over my years of gardening experience, I've learned that effective temperature management can make or break a greenhouse.

When you first set up your greenhouse, understanding the basic principles of temperature control is essential. Plants generally have two temperature ranges to consider: the optimal range and the stress range. The optimal range is where plants grow best, while the stress range can lead to poor growth or even plant death if not addressed. For instance, many vegetables thrive between 65-75°F (18-24°C), while tropical plants might prefer 75-85°F (24-29°C).

To maintain these conditions, it's vital to install a reliable heating system. There are several types of greenhouse heaters to choose from. Gas heaters are popular due to their efficiency and ability to warm large spaces. They can be powered by propane or natural gas. Electric heaters are another option, offering precise temperature control and the convenience of being easy to install. For smaller greenhouses, a simple electric heater might suffice, but for larger setups, you might need a more robust system.

Another essential component of temperature control is ventilation. Without proper ventilation, your greenhouse can quickly become a heat trap. Automatic vents are an excellent investment; they open and close based on the internal temperature, helping to regulate the environment.

Roof vents are particularly effective, as hot air rises and can be expelled through the top of the greenhouse. In addition, side vents or louvered vents can be used to increase airflow and prevent overheating.

Shading is another method to control temperature. During the summer months, excessive sunlight can increase the greenhouse temperature significantly. Shade cloths can be placed over your greenhouse to block some of the sun's rays. These come in various densities, allowing you to adjust the level of shading based on your needs. Additionally, external shading systems can be installed to provide shade while still allowing for air circulation.

Monitoring temperature consistently is crucial. Place thermometers at various points within the greenhouse to get a comprehensive view of temperature variations. Digital temperature sensors connected to a climate control system can provide real-time data and alerts if temperatures move outside the desired range. This technology can automate the adjustment of heating or ventilation systems, ensuring a stable environment.

Consider the role of thermal mass in temperature control. Materials like water barrels or concrete can absorb heat during the day and release it at night, helping to moderate temperature fluctuations. This is especially useful for greenhouses with less sophisticated heating systems.

Seasonal adjustments are necessary. During winter, you might need to increase heating to combat the cold, while in summer, focus on ventilation and shading to keep temperatures from rising too high. Regularly check and calibrate your heating and ventilation systems to ensure they are operating efficiently and effectively.

Humidity Management

Managing humidity in your greenhouse is as crucial as controlling temperature. Plants release moisture into the air through transpiration, and if this moisture accumulates too much, it can lead to problems like mold, mildew, and other diseases. Conversely, low humidity can lead to dehydration and hinder plant growth. Achieving the right balance is key to a thriving greenhouse.

To start, it's essential to understand the specific humidity needs of your plants. Most greenhouse plants thrive in a humidity range of 50-70%. Some tropical plants might require higher humidity, while succulents and cacti prefer lower levels. Researching the needs of your plant species will guide your humidity management practices.

One of the most common methods to increase humidity is by using humidifiers. These devices add moisture to the air and can be particularly useful during the winter months when indoor heating can significantly lower humidity levels. There are various types of humidifiers, including misting systems and vaporizing units. Misting systems release fine droplets of water into the air while vaporizing units use heat to create steam.

Conversely, if you need to reduce humidity, dehumidifiers are your best bet. They work by removing excess moisture from the air, which can help prevent fungal infections and maintain plant health. Dehumidifiers come in different sizes and capacities, so choose one that suits the size of your greenhouse and the extent of humidity control needed.

Ventilation is another crucial aspect of humidity management. Proper airflow helps to prevent the buildup of moisture and reduces the risk of disease. Fans can be used to promote air circulation, ensuring that humidity is evenly distributed throughout the greenhouse. Exhaust fans can also be installed to expel humid air and replace it with drier air from outside.

Monitoring humidity levels is vital for effective management. Hygrometers are simple instruments that measure relative humidity, and placing them at different points in your greenhouse can give you a clear picture of moisture levels. Digital humidity sensors are available that can provide real-time data and integrate with climate control systems for automated adjustments.

Consider the impact of your watering practices on humidity. Overwatering can increase humidity levels, so adjust watering schedules according to your greenhouse conditions.

Seasonal changes also affect humidity. During the summer, increased transpiration and higher temperatures can lead to higher humidity levels. Conversely, winter's dry air can reduce humidity, necessitating the use of humidifiers. Regularly adjust your humidity management strategies to align with the changing seasons and your plants' needs.

Light Requirements

Light is a fundamental factor in greenhouse gardening, influencing everything from seed germination to flowering and fruiting. Over my years of gardening, I've learned that understanding and managing light can make a significant difference in your greenhouse's success.

Natural sunlight is the primary light source for greenhouses, but during periods of low light or when growing plants that require specific light conditions, supplemental lighting becomes crucial. The intensity and duration of light impact plant growth.

Most plants have specific light requirements, typically measured in terms of light intensity and duration. For example, seedlings and young plants often require longer periods of light, around 16 hours per day, to promote strong growth. Mature plants may need around 12 hours of light, depending on their growth stage and type.

Grow lights are used to supplement natural sunlight and provide the necessary light spectrum for plant growth. There are several types of grow lights available, including fluorescent, high-intensity discharge (HID), and light-emitting diode (LED) lights. Fluorescent lights are commonly used for seedlings and young plants, as they provide a balanced light spectrum and are energy-efficient.

LED grow lights have become increasingly popular due to their energy efficiency and ability to provide a customizable light spectrum. They can be adjusted to emit different wavelengths of light, catering to various growth stages and plant types. For example, blue light promotes vegetative growth, while red light supports flowering and fruiting.

Lights should be positioned at an appropriate distance from the plants to avoid light burn or insufficient light. Generally, lights should be placed 12-24 inches above the plants, depending on their intensity and the needs of your plants. Using adjustable hangers can make it easier to modify the height as plants grow.

Timing is another important aspect of light management. Timers can be used to automate the light schedule, ensuring that plants receive consistent and appropriate light exposure. This automation helps to mimic natural day-night cycles and supports healthy plant development.

Shading may also be necessary to prevent overheating and excessive light intensity during peak sunlight hours.

Shade cloths can be used to diffuse sunlight and reduce the risk of plant stress or sunburn. These cloths come in different densities, allowing you to control the amount of light that enters the greenhouse.

Regularly monitor light conditions within your greenhouse. Light meters can be used to measure light intensity and ensure that your plants are receiving the appropriate amount of light. Adjust your lighting setup based on these measurements and the growth stage of your plants to optimize their health and productivity.

Adjusting for Seasonal Changes

Seasonal changes bring a variety of challenges to greenhouse management. As the weather outside shifts, so do the conditions within your greenhouse. Adapting to these changes is crucial for maintaining a stable environment that supports plant health and productivity. Over the years, I've found that understanding and adjusting for seasonal changes can make a significant difference in the success of your greenhouse.

In the summer, one of the primary concerns is managing high temperatures. As the sun's intensity increases, your greenhouse can quickly become a heat trap. To combat this, proper ventilation is key. Automatic roof vents and side vents can help to expel hot air and replace it with cooler air from outside. Additionally, using shade cloths can reduce the amount of direct sunlight entering the greenhouse, helping to prevent overheating.

During summer, humidity levels can also rise due to increased transpiration from your plants. Proper ventilation and the use of dehumidifiers can help manage excess moisture in the air. Regularly check humidity levels and adjust your systems as needed to prevent issues such as mold and mildew.

As fall approaches and temperatures begin to drop, you'll need to prepare for cooler conditions. This often involves adjusting your heating system to maintain a consistent temperature. Depending on the severity of the temperature drop, you may need to increase heating or add additional heaters to ensure that your greenhouse remains warm.

In winter, reduced daylight and lower temperatures can present challenges. Supplemental lighting may be necessary to provide plants with adequate light during shorter days. Grow lights with adjustable spectrums can help mimic natural sunlight and support plant growth. Additionally, using heaters to maintain a consistent temperature is crucial, especially during extreme cold spells.

Winter also brings lower humidity levels, which can lead to dehydration of your plants. Using humidifiers can help to maintain optimal moisture levels in the air. Monitor humidity regularly and adjust your humidifiers as needed to ensure that your plants receive the necessary moisture.

Spring marks a time of transition, with increasing temperatures and longer daylight hours. It's important to gradually adjust your greenhouse conditions to accommodate these changes. For example, you might need to reduce the use of heaters and increase ventilation as temperatures rise. Additionally, as plants begin to grow more actively, you'll need to ensure that light and humidity levels are adjusted to support their growth.

Throughout the year, regularly assess your greenhouse conditions and make adjustments based on seasonal changes. Keeping a log of temperature, humidity, and light levels can help you identify trends and prepare for upcoming seasons. By proactively managing these changes, you can create a stable and thriving environment for your plants year-round.

CHAPTER 7

Ventilation and Air Circulation

"A greenhouse's success depends on more than just sunlight; it thrives on the invisible currents of air that keep the environment balanced and the plants healthy."

Ventilation and air circulation are often overlooked elements in greenhouse management, yet they are essential for creating a stable and healthy growing environment. Without proper ventilation, your plants can suffer from extreme temperatures, high humidity, and poor air quality, all of which can stifle growth and promote disease.

As someone who has spent years optimizing greenhouse environments, I've learned that good air circulation is as crucial as water and nutrients.

Importance of Proper Ventilation

Ventilation is the cornerstone of a healthy greenhouse environment. It serves multiple purposes, including temperature regulation, humidity control, CO2 replenishment, and air circulation. Without proper ventilation, a greenhouse can quickly become inhospitable to plants, leading to heat stress, disease proliferation, and stunted growth. Let's dive deeper into why each of these functions is vital for your greenhouse.

Temperature Regulation: The primary function of ventilation is to regulate temperature. Greenhouses are designed to trap heat, utilizing the greenhouse effect to create a warm, nurturing environment for plants. However, without adequate ventilation, temperatures can rise to levels that are detrimental to plant health.

For example, many plants thrive within a specific temperature range, typically between 18-30°C (64-86°F). Exceeding this range can cause physiological stress, leading to wilting, leaf scorch, and in severe cases, plant death. Proper ventilation allows for the exchange of hot air with cooler air, maintaining a stable and optimal temperature range.

Humidity Control: As plants transpire, they release water vapor into the air, increasing the humidity levels inside the greenhouse. While a certain level of humidity is beneficial, promoting healthy growth and preventing dehydration, excessive humidity can be problematic.

High humidity can create a conducive environment for fungal diseases, such as powdery mildew and botrytis, which can devastate crops. It can also hinder pollination in some plants and reduce the efficiency of photosynthesis. Ventilation helps to dissipate excess moisture, keeping humidity levels in check and preventing these issues from arising.

CO2 Replenishment: Carbon dioxide (CO2) is a critical component of photosynthesis, the process by which plants convert light energy into chemical energy. In a closed greenhouse environment, CO2 can be rapidly depleted, especially during periods of intense photosynthetic activity. Without adequate levels of CO2, photosynthesis slows down, stunting plant growth and reducing yield.

Proper ventilation ensures a continuous supply of fresh air, rich in CO2, or growers may supplement CO2 levels to enhance growth, but this practice requires careful management of ventilation to avoid CO2 depletion.

Air Circulation: Good ventilation promotes uniform air circulation, preventing the formation of microclimates within the greenhouse. Microclimates are small areas where temperature, humidity, and CO2 levels differ significantly from the surrounding environment. These variations can lead to inconsistent plant growth and increased susceptibility to pests and diseases. Proper ventilation ensures a consistent environment throughout the greenhouse, providing all plants with equal access to optimal growing conditions. This uniformity is crucial for maintaining the health and productivity of your crops.

Ventilation Systems

Greenhouse ventilation systems can be broadly categorized into two types: passive (natural) ventilation and active (mechanical) ventilation. Each system has its advantages and considerations, depending on the size of the greenhouse, the climate, and specific plant requirements.

Passive Ventilation

Passive ventilation relies on natural air movement to ventilate the greenhouse. This system typically involves the use of vents, louvers, and openings strategically placed to facilitate airflow. The basic principle of passive ventilation is that warm air rises. Roof vents allow hot air to escape, while cooler air enters through side vents or louvers at the lower levels. This natural airflow creates a convection current that helps to regulate temperature and humidity.

One of the significant benefits of passive ventilation is its energy efficiency. Since it doesn't require electricity, it is cost-effective and environmentally friendly. However, the effectiveness of passive ventilation can be influenced by external weather conditions. For instance, on calm days with little wind, the natural airflow may not be sufficient to achieve the desired ventilation. Additionally, passive systems may struggle to maintain optimal conditions in extreme climates, where temperature and humidity levels can vary widely.

To maximize the efficiency of passive ventilation, it's essential to design the greenhouse with appropriate vent placement and size. The size of the vents should be proportional to the size of the greenhouse to ensure adequate air exchange. In some cases, passive ventilation can be enhanced with the addition of thermal screens or shading systems to reduce heat gain and further control the internal environment.

Active Ventilation

Active ventilation uses mechanical systems, such as fans and blowers, to move air in and out of the greenhouse. This method provides more precise control over the greenhouse environment, making it suitable for larger structures or those in regions with extreme temperature fluctuations. Active systems can include exhaust fans, intake fans, and horizontal air flow (HAF) fans.

Exhaust fans are typically installed near the roof to expel hot air, while intake fans at the opposite end bring in fresh, cooler air from outside. HAF fans are used to circulate air horizontally within the greenhouse, ensuring uniform temperature and humidity distribution. The primary advantage of active ventilation is its reliability and effectiveness, regardless of external weather conditions. It allows for consistent air exchange and climate control, beneficial for sensitive crops or in commercial operations where uniformity and predictability are crucial.

However, active systems come with higher initial costs for equipment and ongoing operational costs due to electricity usage. It's also essential to maintain these systems regularly to prevent malfunctions and ensure their efficiency. Despite these considerations, the benefits of precise environmental control often outweigh the costs, especially in high-value crop production.

Hybrid Systems

In some cases, a combination of passive and active systems, known as hybrid ventilation, can be used. This approach leverages the benefits of both methods, using passive systems to reduce energy consumption and active systems for additional control when needed. For example, passive vents can be opened during mild weather, while fans can be activated during hot or humid conditions to provide additional cooling and air exchange.

Installation Tips

Installing an effective ventilation system in your greenhouse is crucial for maintaining an optimal growing environment. Here are some essential tips and considerations to ensure proper installation and functionality.

Site Assessment and Design: Before installing any ventilation system, assess the specific characteristics of your greenhouse site. Consider factors such as prevailing wind direction, surrounding landscape, and potential obstructions. These factors can influence the effectiveness of both passive and active ventilation systems. For instance, placing intake vents on the windward side can enhance natural airflow in passive systems. In contrast, active systems should be positioned to avoid obstructions that could disrupt air circulation.

Vent Placement and Size: The placement and size of vents are critical for efficient air exchange. For passive systems, roof vents should be installed at the highest point to allow hot air to escape, while side vents or louvers should be positioned at lower levels to facilitate the intake of cooler air. Ensure that vents are large enough to provide adequate ventilation for the size of your greenhouse. A general guideline is to have ventilation openings that constitute at least 20% of the greenhouse floor area. This proportion can vary based on the specific design and environmental conditions.

Fan Selection and Positioning: For active ventilation systems, selecting the right type and size of fans is crucial. The capacity of the fans should be matched to the volume of air in the greenhouse, ensuring efficient air exchange. Exhaust fans should be placed near the ceiling to expel hot air, while intake fans should be positioned at the opposite end to draw in fresh air. It's essential to avoid placing fans in positions that create direct drafts on plants, as this can cause stress and damage. Aim for a uniform distribution of air throughout the greenhouse.

Consider Automation: Automation can significantly enhance the efficiency and convenience of your ventilation system. Automated vents, fans, and environmental control systems can be programmed to respond to specific temperature, humidity, and CO2 levels. This allows for precise and consistent climate control, even when you're not present. Automation can also help reduce energy consumption by optimizing the operation of fans and other equipment based on real-time conditions.

Accessibility for Maintenance: When installing ventilation systems, ensure that all components are easily accessible for regular maintenance and adjustments. This includes vents, fans, sensors, and control systems. Regular maintenance is essential to keep the system running efficiently and to extend its lifespan. Accessibility will also facilitate quick responses to any issues that may arise, minimizing disruptions to your greenhouse environment.

Electrical Considerations: For active ventilation systems, consider the electrical requirements and ensure that your greenhouse has the necessary power supply and safety features. This includes proper wiring, circuit breakers, and grounding. It's advisable to consult with a qualified electrician to ensure that the installation meets all safety standards and codes.

Managing Air Circulation

Air circulation within a greenhouse is a critical aspect of maintaining a stable and healthy growing environment. Proper air circulation helps distribute temperature, humidity, and CO2 evenly throughout the greenhouse, preventing the formation of microclimates and ensuring consistent conditions for all plants. In this section, we'll explore the key components of managing air circulation and the strategies you can implement to optimize it.

Horizontal Air Flow (HAF) Fans: Unlike exhaust and intake fans, which facilitate air exchange with the outside, HAF fans are designed to move air horizontally within the greenhouse. This circulation helps to mix the air, preventing stratification and ensuring that all areas of the greenhouse receive uniform conditions.

When installing HAF fans, it's important to position them strategically to create a continuous, circular airflow pattern. This pattern helps to eliminate dead spots where air movement is minimal and conditions can become stagnant. Ideally, the fans should be placed above the plant canopy and angled slightly downward to encourage air movement at plant level.

Proper Spacing and Layout: Overcrowding plants can restrict airflow, creating areas of high humidity and low CO2 levels. To prevent this, ensure that plants are spaced adequately, allowing for free movement of air between and around them. This not only promotes better air circulation but also reduces the risk of disease spread and pest infestations.

In addition to plant spacing, consider the placement of other equipment and structures within the greenhouse. Large objects, such as benches, shelving, or equipment, can obstruct airflow and create microclimates. Arrange these elements in a way that promotes unobstructed air movement throughout the greenhouse.

Temperature and Humidity Monitoring: Regular monitoring of temperature and humidity levels is essential for managing air circulation effectively. Use sensors and data loggers to track these parameters in different areas of the greenhouse. This data can help you identify any areas with significant variations, indicating poor air circulation or the presence of microclimates.

Based on the monitoring data, you can make adjustments to your ventilation and air circulation systems. For example, if you notice high humidity levels in certain areas, you may need to increase airflow or adjust the positioning of fans. Similarly, if temperature variations are detected, you may need to enhance ventilation or shading to achieve more uniform conditions.

Automated Systems: Automating your ventilation and air circulation systems can greatly enhance their efficiency and effectiveness. Automated systems use sensors to monitor environmental conditions in real time and adjust fans, vents, and other equipment accordingly. This allows for precise control of temperature, humidity, and CO2 levels, ensuring consistent conditions throughout the greenhouse.

Automated systems can also help optimize energy usage by operating only when necessary. For example, fans can be programmed to run at different speeds or turn off when optimal conditions are achieved. This not only reduces energy costs but also extends the lifespan of the equipment.

Regular Maintenance and Adjustments: Regular maintenance of air circulation equipment is crucial for ensuring their optimal performance. Clean fans, vents, and other components regularly to prevent the buildup of dust and debris, which can reduce airflow and efficiency. Check for any signs of wear or damage, such as loose connections or worn-out fan blades, and address these issues promptly.

In addition to maintenance, make periodic adjustments to your air circulation system based on seasonal changes and plant growth stages.

Maintaining Ventilation Systems

Maintaining ventilation systems in a greenhouse is essential for ensuring their long-term efficiency and effectiveness. Proper maintenance not only extends the lifespan of the equipment but also helps to maintain a stable and healthy growing environment for your plants. In this section, we'll cover the key maintenance tasks and best practices for keeping your ventilation systems in top condition.

Cleaning Vents and Fans: Dust, dirt, and debris can accumulate on vents, fans, and other components of your ventilation system, reducing their efficiency and effectiveness. Regular cleaning is essential to ensure optimal airflow and prevent blockages. For passive systems, clean the vents and louvers regularly to remove any buildup. For active systems, clean the fan blades, grilles, and housing to prevent dust accumulation.

Use a soft brush or cloth to clean the components, being careful not to damage any delicate parts. For more thorough cleaning, you can use a mild detergent and water solution. Be sure to disconnect any electrical components before cleaning to avoid the risk of electrical shock. After cleaning, inspect the components for any signs of wear or damage and replace them if necessary.

Inspecting Electrical Components: For active ventilation systems, regular inspection of electrical components is crucial. Check the wiring, switches, and connections for any signs of wear, corrosion, or damage. Loose or damaged wiring can pose a safety hazard and reduce the efficiency of the system. If you notice any issues, consult a qualified electrician to repair or replace the components as needed.

In addition to wiring, inspect the motor and other mechanical parts of the fans. Listen for any unusual noises, such as grinding or squeaking, which may indicate worn-out bearings or other mechanical issues. Address these problems promptly to prevent further damage and ensure the smooth operation of the system.

Lubricating Moving Parts: Lubrication is essential for maintaining the smooth operation of mechanical components, such as fan bearings. Over time, the lubrication can dry out, leading to increased friction, wear, and noise. Regularly lubricate the moving parts with a suitable lubricant, following the manufacturer's recommendations. Be careful not to over-lubricate, as this can attract dust and debris, leading to further issues.

Testing and Calibrating Sensors: If your ventilation system includes automated controls, such as temperature, humidity, and CO2 sensors, it's important to test and calibrate these sensors regularly. Accurate sensors are essential for maintaining a stable greenhouse environment, as they provide the data needed to adjust the ventilation system. Test the sensors to ensure they are providing accurate readings, and calibrate them if necessary. Follow the manufacturer's instructions for testing and calibration procedures.

Monitoring System Performance: Regularly monitor the performance of your ventilation system to ensure it is operating efficiently. Keep a log of environmental conditions, such as temperature, humidity, and CO2 levels, and compare them to the desired setpoints. If you notice any discrepancies, investigate the cause and make the necessary adjustments.

For active systems, monitor the power consumption of the fans and other electrical components. An increase in power consumption can indicate a problem, such as a mechanical issue or a blockage in the airflow. Address these issues promptly to prevent further damage and reduce energy costs.

Seasonal Adjustments: As the seasons change, the environmental conditions inside the greenhouse can vary significantly. It's important to adjust your ventilation system to accommodate these changes. Adjust the settings of your vents, fans, and automated controls accordingly. Consider installing seasonal accessories, such as shading screens or thermal curtains.

Documentation and Record Keeping: Keeping detailed records of maintenance activities, inspections, and adjustments is important for tracking the performance of your ventilation system. Document the dates of cleaning, inspections, lubrication, and any repairs or replacements. This information can help you identify patterns, plan future maintenance, and make informed decisions about system upgrades or replacements.

Maintaining your greenhouse ventilation system is crucial for ensuring a healthy and productive growing environment. Regular cleaning, inspection, lubrication, and calibration are essential tasks that will keep your system running smoothly and efficiently. A well-maintained ventilation system is an investment in the health and success of plants of your greenhouse.

CHAPTER 8
Heating and Cooling Your Greenhouse

"A well-regulated greenhouse climate is the difference between a garden's potential and its peak." - Unknown

Creating an environment where your plants can thrive all year round requires more than just planting them in the right soil or providing the right amount of water. In a greenhouse, controlling temperature is crucial to ensure that your plants grow healthy and productive. As we dive into the specifics, I'll share insights from my experience, helping you navigate the complexities of greenhouse climate management.

Maintaining Optimal Temperatures

The primary goal of managing a greenhouse's temperature is to create an environment that mimics the ideal growing conditions for your chosen plants. Optimal temperatures vary depending on the type of plants you're cultivating, but generally, most plants thrive in a range between 65°F and 75°F (18°C to 24°C). However, achieving this is difficult in areas with extreme climates.

To maintain these optimal conditions, it's important to consider both the day and night temperatures. Daytime temperatures can rise significantly, especially when the sun is shining brightly. In contrast, nighttime temperatures can drop, sometimes drastically. This fluctuation can stress plants, affecting their growth and productivity. Therefore, effective temperature management requires a balanced approach to both heating and cooling.

A significant factor in achieving optimal temperatures is understanding the thermal mass of your greenhouse, which refers to the ability of materials to absorb, store, and release heat. Materials such as stone, water barrels, or concrete floors can help with this, as they absorb heat during the day and slowly release it at night, thereby moderating temperature swings. This is particularly useful for preventing the greenhouse from becoming too cold at night and too hot during the day.

Another important aspect to consider is the insulation of your greenhouse, which helps maintain stable temperatures by minimizing heat exchange with the outside environment.

Ventilation also plays a critical role in temperature management. While it primarily helps in cooling, controlled ventilation can also prevent heat buildup during the winter.

For instance, during sunny winter days, temperatures inside the greenhouse can rise quickly. In such cases vents prevent overheating.

Heating Options

When it comes to heating your greenhouse, there are several options available, each with its own advantages and disadvantages. The choice depends on your greenhouse size, budget, and the local climate. One of the most traditional methods is the use of electric or gas heaters.

These heaters can quickly raise the temperature inside the greenhouse, making them ideal for cold climates where maintaining warmth is a constant challenge.

Electric heaters are generally easy to install and can be controlled with thermostats for precise temperature regulation. They are particularly suitable for small to medium-sized greenhouses. However, they can be costly to operate, especially in regions where electricity rates are high. Gas heaters, on the other hand, can be more cost-effective for larger greenhouses. They require proper ventilation to prevent the buildup of harmful gases like carbon monoxide, making safety considerations paramount.

Another heating option is radiant heating, which involves installing systems that emit infrared radiation. These systems can be embedded in the greenhouse floor or installed overhead. Radiant heating warms the plants and soil directly, rather than the air, providing a more uniform temperature distribution. This method is energy-efficient and can reduce heating costs over time. However, the initial installation can be expensive, and it requires careful planning and design.

For those looking for a more sustainable option, solar heating is an excellent choice. By harnessing the sun's energy, solar panels can provide a renewable source of heat.

While the initial investment can be high, solar heating systems can significantly reduce energy costs in the long run. They are particularly effective in regions with ample sunlight, although supplementary heating may still be needed during prolonged cloudy periods or at night.

Passive solar heating is another method worth considering. It involves designing the greenhouse in such a way that it maximizes solar gain during the day and minimizes heat loss at night. This can be achieved through proper orientation of the greenhouse, the use of thermal mass, and adequate insulation.

For example, a greenhouse oriented with its long axis running east-west will have maximum exposure to the sun, optimizing heat gain. This design, combined with thermal mass elements, can significantly reduce the need for artificial heating.

Cooling Techniques

Cooling a greenhouse is equally important as heating, particularly during hot summer months. Without proper cooling, temperatures can soar, leading to heat stress in plants, which can stunt growth and reduce yields. There are several techniques to cool a greenhouse, each suitable for different climates and setups.

One of the most common cooling methods is natural ventilation. This involves using roof vents, side vents, and louvers to allow hot air to escape and cooler air to enter. The effectiveness of natural ventilation depends on the design of the greenhouse and the external weather conditions. In climates with a consistent breeze, natural ventilation can be highly effective. However, it may not be sufficient in very hot or humid conditions, where additional cooling methods are needed.

Shade cloths are another practical cooling technique. These cloths can be draped over the greenhouse structure to reduce the amount of sunlight entering the space, thereby lowering temperatures. Shade cloths come in different densities, allowing you to control the amount of light and heat that penetrates the greenhouse.

This method is particularly useful for protecting plants that are sensitive to intense sunlight. However, it can also reduce the amount of light available for photosynthesis, so it's essential to strike a balance.

For more controlled cooling, evaporative cooling systems are a popular choice. These systems use fans and water-saturated pads to cool the air. As air passes through the pads, it is cooled by the evaporation of water, then circulated throughout the greenhouse.

This method is especially effective in dry climates, as it adds humidity to the air while lowering the temperature. However, in already humid environments, evaporative cooling may not be as efficient, and careful management of humidity levels is necessary to prevent disease.

Misting systems can also be used to cool a greenhouse. By spraying a fine mist of water into the air, these systems reduce the air temperature through evaporation. Misting not only cools the air but also increases humidity, which can be beneficial for certain plant species. However, it's important to monitor humidity levels closely, as excessive moisture can promote the growth of mold and mildew.

Insulation Strategies

Insulation plays a vital role in both heating and cooling. By minimizing heat loss during the winter and reducing heat gain during the summer, proper insulation can help maintain a stable internal temperature and reduce energy costs.

One of the most effective ways to insulate a greenhouse is by using double-layered glazing. This involves having two layers of material, such as polycarbonate or glass, with an air gap in between. The air gap acts as an insulating barrier, reducing heat transfer.

Another insulation strategy is to use thermal screens or curtains. These can be drawn across the inside of the greenhouse at night to retain heat and opened during the day to allow light and warmth in.

Thermal screens are particularly useful in large greenhouses, where the cost of heating can be significant. They also provide an additional layer of shading during hot days, helping to keep the temperature down.

In addition to these methods, insulating the greenhouse foundation and walls can prevent heat loss. Materials such as foam boards can be installed around the perimeter of the greenhouse to reduce thermal bridging—the transfer of heat through the structure. This is especially important in colder climates, where the ground itself can act as a heat sink, drawing warmth away from the greenhouse.

Seasonal Temperature Management

Managing temperatures in a greenhouse is not a static task; it requires adjustment throughout the year as seasons change. In winter, the focus is on retaining heat and preventing cold drafts. This may involve closing vents and using supplementary heating during nighttime and cloudy days. Regular monitoring of temperature and humidity levels is crucial to ensure that the environment remains conducive to plant growth.

During the summer, the emphasis shifts to cooling. Increased ventilation, shading, and possibly even misting systems may be needed to combat the heat. It's also important to monitor the internal humidity levels, as high humidity can lead to the growth of mold and mildew. Using a combination of fans and dehumidifiers can help manage moisture levels effectively.

In transitional seasons, such as spring and fall, maintaining the right temperature can be particularly challenging due to unpredictable weather patterns. During these times, a combination of passive and active methods may be needed. For instance, on a sunny spring day, you might need to open vents to prevent overheating, while at night, you may need to draw thermal curtains to retain warmth.

Having a reliable climate control system that automates heating, cooling, and ventilation can be a game-changer. Modern systems can be programmed to maintain desired temperature and humidity levels, adjusting automatically based on real-time conditions. They help grow healthier plants and better yields.

Energy Efficiency Tips

Energy efficiency is not just about saving money; it's also about minimizing the environmental impact of your greenhouse. One of the simplest ways to improve energy efficiency is by optimizing the use of natural resources. For instance, positioning your greenhouse to maximize sunlight exposure reduces the need for artificial lighting and heating. Similarly, using thermal mass to store heat can reduce reliance on electric or gas heaters.

Another tip is to regularly maintain your greenhouse equipment. Dirty or malfunctioning equipment can consume more energy than necessary, leading to higher costs and reduced effectiveness. Additionally, sealing any gaps or cracks in the greenhouse structure can prevent heat loss and reduce the need for supplemental heating.

Solar panels can be used to power ventilation systems, heaters, or even lighting. Using LED grow lights reduces energy consumption and produce less heat than traditional light options. This reduces the cooling load on your greenhouse. They also have a longer lifespan, which means fewer replacements and less waste.

In conclusion, managing the temperature in your greenhouse is a multifaceted task that requires a combination of heating, cooling, insulation, and energy efficiency strategies.

CHAPTER 9

Setting Up Your Greenhouse for Success

"Success in gardening comes from preparation and foresight—plan your garden, and your plants will thrive."

The moment you've been waiting for is here: setting up your greenhouse and preparing it for the first batch of plants. This chapter is all about laying the groundwork to ensure your greenhouse becomes a thriving, productive space.

From choosing the right growing medium to organizing planting areas, these initial steps are crucial for creating an environment that supports healthy plant growth.

With decades of gardening experience under my belt, I'll guide you through each aspect, sharing practical tips and insights to help you get started on the right foot.

Choosing the Right-Growing Medium

The growing medium you choose is fundamental to your greenhouse's success, as it provides the nutrients, air, and water your plants need to thrive. Unlike traditional outdoor gardening, where soil is often used, greenhouse gardening offers more options, including soilless mixes. Each medium has its advantages and disadvantages, and the best choice depends on your plant's specific needs and your gardening style.

For many greenhouse gardeners, soilless mixes are the preferred choice because they offer superior control over nutrient content and water retention. These mixes often consist of materials like peat moss, perlite, vermiculite, and coconut coir. Peat moss is excellent for water retention and provides an acidic environment, which is ideal for acid-loving plants. Perlite and vermiculite, on the other hand, are lightweight and help with aeration and drainage, preventing root rot. Coconut coir is a sustainable alternative to peat moss and offers excellent water retention and aeration properties.

One of the key benefits of soilless mixes is the reduced risk of soil-borne diseases and pests. Since these mixes are typically sterile, they provide a clean slate for growing plants, minimizing the chances of disease outbreaks. However, because they contain no natural nutrients, you'll need to supplement them with fertilizers. This allows for precise control over the nutrients your plants receive, which can be particularly advantageous for growing specialized crops.

If you prefer using soil, choosing a high-quality, well-draining potting mix is essential. Garden soil alone is usually too heavy and compact for greenhouse use, leading to poor drainage and aeration. A good potting mix should be light, airy, and rich in organic matter. You can either purchase a pre-made mix or create your own by blending garden soil with compost, sand, and other amendments. Adding compost not only improves soil structure but also provides a slow-release source of nutrients.

For those interested in hydroponics, the growing medium takes on a different meaning. In hydroponic systems, plants are grown in a nutrient-rich water solution, with the medium primarily serving as a support system for the plant roots. Common hydroponic mediums include rock wool, clay pellets, and foam. Each of these mediums has unique properties, such as water retention or pH neutrality, which can be used in specific hydroponic systems and plant types.

Regardless of the medium you choose, it's essential to maintain a balance of water, air, and nutrients. Overwatering is a common issue in greenhouse gardening, leading to root rot and other problems. Proper drainage and aeration are critical, especially in a controlled environment where natural air movement is limited.

Regularly check the moisture level of your medium and adjust watering schedules as needed.

Selecting Containers

Containers play a vital role in greenhouse gardening, as they determine the growth potential of your plants and influence the ease of maintenance. The choice of containers depends on the type of plants you're growing, the available space, and your watering habits. While there are many types of containers available, including plastic pots, clay pots, grow bags, and trays, each has its pros and cons.

Plastic pots are popular due to their affordability, lightweight nature, and durability. They are available in various sizes and shapes, making them versatile for different plants.

One of the main advantages of plastic pots is their excellent water retention, which is beneficial for plants that require consistent moisture. However, they can also lead to waterlogging if not adequately drained. To avoid this, choose pots with drainage holes and consider placing a layer of gravel at the bottom to improve drainage.

Clay pots, also known as terracotta pots, are another common choice. They are porous, allowing for better air exchange and preventing soil from becoming waterlogged. This is particularly beneficial for plants that prefer drier conditions, such as succulents and cacti.

However, clay pots are heavier and more prone to breaking than plastic pots. They also tend to dry out faster, requiring more frequent watering. Despite these drawbacks, many gardeners prefer clay pots for their aesthetic appeal and natural cooling effect on the soil.

Grow bags are an excellent option for larger plants or those with extensive root systems. Made from breathable fabric, grow bags promote air pruning of roots, preventing them from becoming root-bound. This leads to healthier root systems and more vigorous plant growth. Grow bags are also lightweight and easy to move, making them ideal for rearranging plants as needed. However, they require more frequent watering due to their increased breathability.

For seedlings and young plants, seed trays and plug trays are highly recommended. These trays allow you to start multiple plants in a small space, making it easy to manage and monitor growth. Once the seedlings are large enough, they can be transplanted into larger containers. Trays are particularly useful for starting plants that will eventually be moved outdoors or into larger greenhouse beds.

No matter what type of container you choose, ensure it provides adequate drainage and is the right size for the plant. As plants grow, their roots expand, requiring more space. Overcrowded roots can lead to stunted growth and reduced yields. When selecting a container, consider the mature size of the plant and choose a pot that allows room for growth. It's also important to sanitize containers before use, especially if they have been used previously, to prevent the spread of diseases.

Essential Gardening Tools

Equipping your greenhouse with the right tools is crucial for efficient and successful gardening. The tools you need will depend on the size of your greenhouse, the types of plants you're growing, and your gardening practices. However, there are some essential tools that every greenhouse gardener should have.

A good pair of gloves is indispensable. They protect your hands from dirt, cuts, and chemicals. Choose gloves that are comfortable, durable, and suitable for the tasks you'll be performing. For delicate tasks like transplanting seedlings, lightweight gloves with good tactile sensitivity are ideal. For heavier tasks, such as handling thorny plants or working with soil, thicker gloves are recommended.

A watering can or hose with a fine nozzle is essential for controlled watering. In a greenhouse, where space is often limited, the ability to water plants precisely without splashing is crucial. A watering can with a long spout allows you to reach plants at the back of the greenhouse without disturbing the soil or foliage.

For larger greenhouses, a hose with a multi-pattern nozzle can provide flexibility in watering, from a gentle mist for delicate seedlings to a pressurized flow for established plants.

Pruning shears are another must-have tool. Regular pruning promotes healthy growth, removes dead or diseased parts, and helps shape the plants. Choose shears that are sharp, comfortable to use, and appropriate for the type of plants you're working with. For smaller plants and detailed work, bypass pruners are ideal. For thicker branches, loppers or pruning saws may be necessary.

A hand trowel is invaluable for tasks such as transplanting seedlings, mixing soil, and planting. Look for a trowel with a comfortable grip and a sturdy blade. A narrow-bladed trowel is useful for working in tight spaces, while a broader blade is better for moving larger amounts of soil.

Other essential tools include a soil thermometer, which helps monitor soil temperature; a pH meter, which measures the acidity or alkalinity of your growing medium; and a hand fork, which is useful for loosening soil and removing weeds. A good quality rake is also beneficial for leveling soil and collecting debris.

In addition to these tools, consider investing in a bench or table for potting and a set of shelves for organizing supplies and tools.

Preparing the Greenhouse Interior

Setting up the interior of your greenhouse is an exciting step, as it involves creating a space that is both functional and aesthetically pleasing. The layout should facilitate easy movement without stepping on plants, efficient use of space, and access to all plants. Start by deciding on the main pathways and work areas. Benches and shelves are essential components of the greenhouse interior. They provide space for potted plants, seed trays, and tools. When setting up benches, consider the height and depth. They should be at a comfortable working height and deep enough to hold the largest pots or trays you plan to use. Shelves can be installed above benches for additional storage or growing space. Use wire or mesh shelves to allow for proper air circulation and light penetration.

For larger plants or those with extensive root systems, consider using raised beds. Raised beds allow for better control of soil quality and drainage. They can be constructed from various materials, including wood, metal, or plastic. When building raised beds, ensure they are deep enough to accommodate the root systems of the plants you intend to grow.

Organize your plants based on their light and temperature requirements. Position sun-loving plants near the front or sides of the greenhouse, where they will receive the most light. Shade-tolerant plants can be placed towards the back or in areas where they are protected from direct sunlight.

Grouping plants with similar needs together simplifies watering and care routines.

Incorporate a watering system that suits your setup and plant needs. Options include drip irrigation, overhead sprinklers, or manual watering with a can or hose. Drip irrigation is particularly efficient, as it delivers water directly to the base of each plant, minimizing waste and reducing the risk of fungal diseases caused by wet foliage.

Consider the aesthetic aspects of your greenhouse interior. Adding decorative elements, such as potted flowers, hanging baskets, or ornamental plants, can enhance the visual appeal of your greenhouse. Personal touches, like garden art or a comfortable chair, can make the space more inviting and enjoyable to spend time in.

Organizing Planting Areas

Efficient organization of planting areas is key to maximizing the productivity and functionality of your greenhouse. Proper planning ensures that each plant receives the care and conditions it needs to thrive. Start by mapping out the layout of your planting areas, taking into account the size and growth habits of your plants.

Divide the greenhouse into zones based on plant needs. For example, designate separate areas for seedlings, mature plants, and specialized crops like orchids or succulents. This zoning allows you to tailor care practices, such as watering and fertilization, to the specific requirements of each group. It also makes it easier to monitor the health and progress of your plants.

Consider the vertical space in your greenhouse. Using vertical growing systems, such as shelves, trellises, or hanging baskets, can significantly increase your planting capacity. Vertical gardening is particularly useful for vining plants like tomatoes or cucumbers, which can be trained to grow upwards, saving valuable ground space. Hanging baskets are ideal for trailing plants or herbs, while tiered shelving can accommodate multiple levels of potted plants.

Label your plants clearly and accurately. Labels can include the plant name, variety, and any special care instructions. This is especially important in a greenhouse with a diverse collection of plants. Proper labeling helps you keep track of planting dates, growth stages, and any treatments or fertilizations applied. Use durable, weather-resistant labels to ensure they remain legible over time.

Create a schedule for routine maintenance tasks, such as watering, pruning, and cleaning. Set up a system for monitoring plant health, checking for signs of pests, diseases, or nutrient deficiencies. Early detection and intervention can prevent minor issues from becoming major problems.

Initial Plant Setup

The initial setup of your plants in the greenhouse is an exciting moment, as it marks the beginning of your gardening journey. Careful planning and preparation during this phase can set the stage for a successful growing season. Start by selecting healthy, vigorous plants or seedlings from a reputable nursery or garden center.

When planting in containers or raised beds, use a high-quality growing medium appropriate for your chosen plants. If using soilless mixes, remember to provide a balanced fertilizer to supply essential nutrients. For soil-based systems, don't forget to add compost or organic matter to improve soil fertility and structure.

Begin by arranging the plants according to their light and temperature needs. Ensure that taller plants do not overshadow shorter ones, which can lead to uneven growth and reduced yields. Leave sufficient space between plants to allow for air circulation and ease of access for maintenance tasks.

Water your plants thoroughly after transplanting to help them establish in their new environment. Use a gentle watering method to avoid disturbing the roots or soil. Monitor the plants closely during the first few weeks, as they may experience transplant shock, characterized by wilting or slowed growth. With proper care, they will quickly recover and begin to thrive.

As your plants grow, provide support as needed. This may include stakes, trellises, or cages for climbing plants or spacers for preventing sprawling plants from encroaching on their neighbors.

Also, keep a gardening journal or log to track your activities and observations. Note the planting dates, growth progress, and any issues that arise. This record will be invaluable for future reference, helping you refine your techniques and improve your greenhouse gardening skills.

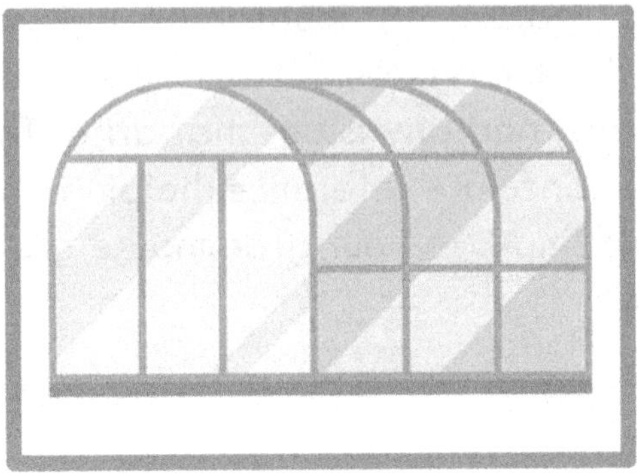

PART 4: GROWING PLANTS IN YOUR GREENHOUSE

CHAPTER 10

Selecting Plants for Your Greenhouse

"A garden is a reflection of the gardener's soul; the choice of plants reveals the essence of one's dreams."

Selecting the right plants for your greenhouse is akin to choosing the perfect palette for an artist's masterpiece. The plants you grow will not only define the aesthetic appeal of your space but also determine the productivity and sustainability of your gardening efforts.

Whether you're aiming to cultivate a bountiful vegetable harvest, a lush ornamental display, or a specialized collection of exotic species, understanding the key factors in plant selection is crucial.

Plant Selection Criteria to Consider

When choosing plants for your greenhouse, several factors must be taken into account to ensure their success. One of the primary considerations is the climate and environmental conditions within your greenhouse.

Greenhouses provide a controlled environment, but the conditions can vary significantly depending on the type of structure, available heating and cooling systems, and local climate. It's essential to select plants that can thrive in the temperature and humidity levels your greenhouse can maintain.

The intended purpose of your greenhouse influences plant selection. Are you growing for personal enjoyment, commercial production, or educational purposes? Your goals will determine the types of plants you choose and the level of care they require. For instance, a greenhouse dedicated to vegetable production will prioritize high-yield, fast-growing crops, while an ornamental greenhouse might focus on rare and beautiful species.

Space is a practical consideration that should not be overlooked. The size and layout of your greenhouse dictate how many plants you can comfortably grow and the type of plants you can accommodate. Consider the mature size of the plants, including their height, spread, and root depth. Overcrowding can lead to poor air circulation, increased disease risk, and competition for resources, so it's essential to plan for adequate spacing.

Consider your personal preferences and experience level. While it may be tempting to fill your greenhouse with exotic and challenging plants, it's wise to start with species that match your gardening skills. As you gain experience and confidence, you can gradually introduce more complex varieties. Additionally, choosing plants you enjoy and have a passion for will make the gardening process more rewarding and enjoyable.

Companion Planting

Companion planting is a gardening practice that involves growing different plants together to enhance growth, deter pests, and improve soil health. This technique is particularly beneficial in a greenhouse setting, where space is limited and optimizing plant health is crucial. When planning your greenhouse garden, consider pairing plants that complement each other in terms of growth habits, nutrient needs, and pest resistance.

One common example of companion planting is the "Three Sisters" method, traditionally used by Native American tribes. This method involves planting corn, beans, and squash together. The corn provides a natural trellis for the beans to climb; the beans fix nitrogen in the soil, enriching it for the other plants, and the squash spreads along the ground, suppressing weeds and retaining soil moisture. This combination creates a mutually beneficial environment for all three crops.

Another effective pairing is tomatoes and basil. Basil not only enhances the flavor of tomatoes but also repels pests like aphids and whiteflies. Similarly, planting marigolds among vegetables can deter nematodes and other harmful insects due to the natural chemicals they release into the soil. These simple combinations can improve the health and productivity of your greenhouse garden.

It's important to avoid planting certain combinations that can have negative effects. For example, some plants release allelopathic chemicals that inhibit the growth of nearby plants. Black walnut trees are a well-known example, as they produce juglone, which can be toxic to many other plants. Always research the compatibility of the plants you intend to grow together to ensure they support each other's growth.

Companion planting also involves considering the physical layout and spacing of plants. Taller plants can provide shade for shorter, shade-tolerant plants, while spreading plants can act as ground cover, reducing evaporation and weed growth.

By strategically arranging your plants, you can create a smart and productive growing environment.

Selecting Seeds and Seedlings

The foundation of a successful greenhouse garden lies in the quality of the seeds and seedlings you start with. Selecting the right seeds or seedlings is crucial for ensuring healthy growth and high yields. There are several factors to consider when choosing between seeds and seedlings, including the types of plants, growing season, and your level of experience.

Seeds offer the advantage of being more cost-effective and providing a wider variety of plant options. They allow you to start the growing process from scratch, giving you complete control over the plant's growth and development.

However, starting from seeds requires more time and effort, as it involves germination, careful monitoring, and transplanting.

When selecting seeds, check the seed packet for information on germination rates, planting depth, and spacing requirements. It's also important to consider the seed's shelf life and storage conditions. Some seeds, especially those of perishable crops, have a limited viability period and must be used within a specific timeframe.

Seedlings, on the other hand, offer a quicker start to the gardening process, as they have already passed the germination stage. They are ideal for novice gardeners or those with limited time, as they require less initial care and can be transplanted directly into the greenhouse. However, seedlings are generally more expensive than seeds and may have a more limited selection.

When purchasing seedlings, choose healthy, vigorous plants with strong stems and vibrant foliage. Avoid seedlings that appear leggy, discolored, or show signs of pests and diseases. It's also important to acclimatize the seedlings to the greenhouse environment before planting them. This process, known as hardening off, involves gradually exposing the plants to the conditions of the greenhouse, helping them adjust to the temperature and light levels.

Whether you choose seeds or seedlings, proper planting techniques are essential for successful growth. Follow the recommended planting depth and spacing guidelines, and ensure the growing medium is well-prepared and nutrient-rich. Water the plants thoroughly after planting and provide adequate care and monitoring during the initial growth stages.

Seasonal Planting Guides

Greenhouse gardening allows for year-round cultivation, but it's important to consider seasonal variations and how they affect plant growth. Understanding the best times to plant specific crops can help you maximize yields and ensure a continuous supply of fresh produce or flowers. Seasonal planting guides can assist in planning your greenhouse garden, allowing you to take full advantage of the controlled environment.

In the spring, focus on starting warm-season crops that require higher temperatures and longer daylight hours. Tomatoes, peppers, cucumbers, and herbs like basil thrive in the springtime conditions of a greenhouse. This is also an ideal time to start ornamental flowers and tropical plants, as the increasing light levels and temperatures promote healthy growth.

During the summer, the greenhouse can become quite warm, especially if it's not equipped with adequate cooling systems. Use this time to grow heat-loving plants, such as melons, eggplants, and chilies. It's also a good opportunity to experiment with exotic plants that require a consistently warm environment. To manage the heat, consider using shade cloths and ventilation to prevent overheating.

In the fall, as temperatures begin to cool, switch to cool-season crops that prefer lower temperatures. Lettuce, spinach, kale, and radishes are excellent choices for fall planting. These crops can tolerate cooler conditions and often produce better yields when grown in the fall. Fall is also a good time to start bulbs for winter-blooming flowers.

Winter is a unique season for greenhouse gardening, as the greenhouse provides a haven for plants that would otherwise struggle in the cold. Focus on hardy, cold-tolerant crops like carrots, onions, and garlic. Winter greens, such as chard and mustard greens, also thrive in the cooler temperatures of a winter greenhouse. If you have a heated greenhouse, you can extend the growing season for warm-season crops or maintain tropical plants.

Proper planning and organization are key to successful seasonal planting in a greenhouse. Keep track of planting dates, growth stages, and expected harvest times.

Managing Plant Varieties

Managing a diverse collection of plant varieties in your greenhouse can be both rewarding and challenging. The key to success lies in understanding the specific needs of each variety and creating an environment that caters to those needs.

Begin by researching the specific requirements of each plant variety you intend to grow. Factors to consider include light, temperature, humidity, water, and nutrient needs. Group plants with similar requirements together to simplify care and maintenance. For example, place high-light, high-temperature plants in the same area and group shade-tolerant, cool-loving plants in another.

Regularly monitor the health and growth of your plants, paying attention to any signs of stress or disease. Different plant varieties may exhibit different symptoms, so it's important to familiarize yourself with the common issues associated with each type. Early detection and intervention can prevent the spread of pests or diseases and ensure the continued health of your plants.

Crop rotation involves changing the location of specific plant groups each growing season, allowing the soil to recover and reducing the buildup of harmful organisms. For example, after growing a heavy-feeding crop like tomatoes, consider planting a nitrogen-fixing crop like beans in the same area. Rotation also helps prevent the buildup of pests and diseases in the soil, as different plants have different nutrient needs and attract different pests. It also ensures that they receive adequate light and airflow, preventing overcrowding and promoting healthy growth.

Maintain detailed records of your greenhouse activities, including planting dates, growth observations, and any treatments applied. This information will help you refine your gardening practices and improve your understanding of each plant variety. Over time, you'll develop a comprehensive knowledge base that will enable you to manage your greenhouse more effectively.

Finally, enjoy the diversity and beauty of your greenhouse plants. Experiment with new varieties, explore different growing techniques and take pleasure in the unique characteristics of each plant. A well-managed greenhouse filled with a diverse array of plants is not only a source of food and beauty but also a testament to the gardener's skill and dedication.

CHAPTER 11
Seed Starting and Propagation Techniques

"To plant a garden is to believe in tomorrow, but to propagate is to craft the future."

Starting a garden from seeds and propagating plants is a journey of nurturing life from its very inception. It is an intricate and rewarding process that allows gardeners to expand their collections, save on costs, and maintain plant health. We'll explore different propagation methods, the use of greenhouses in the propagation process, and best practices for caring for young plants.

Seed Starting and Propagation Requirements and Process

Starting plants from seeds is one of the most economical and rewarding methods of cultivating a garden. It provides the gardener with a vast selection of plant varieties, often unavailable as mature plants. The process begins with selecting high-quality seeds from reputable sources, ensuring that they are fresh and have a high germination rate.

Different conditions affect the seed starting. Light, temperature, and moisture levels affect the seed starting in different plants. These requirements help in deciding the best environment for seed germination. For instance, some seeds, such as lettuce, need light to germinate and should be sown on the surface of the soil, while others, like beans, need darkness and should be planted deeper.

Understanding these requirements helps create the ideal environment for seed germination.

Growing medium

Choosing a good and fresh growing medium ensures successful seed starting. Mixture of compost, sand vermiculite and peat moss is a great growing medium. A sterile, lightweight seed-starting mix is recommended as it provides excellent drainage and is free of pathogens that could harm young seedlings. Avoid using garden soil, which can be too heavy and may contain weed seeds or diseases. Moistening ensures thorough hydration of the mixture. It adheres seeds better to the soil. Don't use soil direct from the garden for seed starting.

Humidity

Using a humidity dome or covering the trays with plastic wrap can help retain moisture and create a humid environment conducive to germination. Once the seeds have germinated, remove the cover to prevent mold growth and allow for adequate air circulation. Oscillating fan can be used for this purpose.

Lighting

Lighting is crucial for seedling growth. Place the seedlings under grow lights or in a sunny window to ensure they receive sufficient light. Seedlings require about 12-16 hours of light per day, and grow lights can be adjusted to maintain the ideal distance from the plants. If using natural light, rotate the trays regularly to ensure even exposure. Spacing of seeds and depth of planting is done according to the instructions on the packets.

Temperature

Maintain 75°F on average for better seed germination. Optimum temperature and warmth ensure balanced humidity in the greenhouse which prevents fungal diseases. Keep a check on temperature throughout the day. Heat mats are also used to provide the insulation in winters.

Choosing containers

Seed starting trays are reusable clay trays used for plant seedlings not required to plant deeply. They provide easy planting of seeds, from which germinated seedlings can easily be removed without damaging their roots. Plastic trays might be a cheaper option.

Pots are used for seedlings that will grow longer roots upon germination. Clay, plastic, and biodegradable pots are mostly used according to the theme of your greenhouse. Pots have drainage holes for proper drainage and aeration.

Spacing and planting

Plant the seeds according to the instructions on the seed packet, taking care not to overcrowd them. Overcrowding can lead to competition for light and nutrients, resulting in weak seedlings. Cover the seeds with the appropriate amount of soil and gently firm it down to ensure good seed-to-soil contact. After planting, water the seeds lightly using a fine mist to avoid displacing them.

Transplanting

The seedlings are transferred from the containers to larger containers three times the size of the original container or open ground. This usually is done when first one or two leaves appear or when the plant reaches 2 to 3 inches. At this stage, the seedling has established a root system that can provide it nutrition and help it to adapt to a larger container or the open ground.

Not all plants are to be transplanted. Seeds like cucumbers are directly planted in large containers or on the ground.

A cool day is chosen for transplanting the plants for minimum stress.

Propagation Methods

Beyond starting from seeds, plant propagation can also be accomplished through various other methods, including cuttings, division, layering, and grafting. Each method has its own advantages and is suitable for different types of plants.

Cuttings

Cuttings involve taking a portion of a plant, such as a stem, leaf, or root, and encouraging it to grow roots and develop into a new plant. This method is commonly used for plants that do not produce viable seeds or for creating clones of a parent plant. Stem cuttings, for instance, are popular for propagating woody plants and succulents.

To propagate by cuttings, select a healthy parent plant and use a clean, sharp knife or scissors to take a cutting just below a node. Remove any leaves from the lower part of the cutting and dip the cut end in the rooting hormone to encourage root development. Plant the cutting in a moist growing medium and keep it in a warm, humid environment until roots develop.

Division

Division is another effective propagation method, particularly for plants that grow in clumps or have multiple crowns. It involves dividing an existing plant into several smaller sections, each with its own root system. This method is commonly used for perennials like hostas, daylilies, and ornamental grasses.

To propagate by division, dig up the parent plant and carefully separate the clumps, ensuring that each division has a healthy root system. Replant the divisions in the desired location and water them thoroughly.

Layering

Layering is a technique that encourages a plant to produce roots while still attached to the parent plant. This method is particularly useful for plants with long, flexible stems, such as strawberries and blackberries. To propagate by layering, bend a stem down to the ground and cover a portion of it with soil, leaving the tip exposed. The buried section will eventually develop roots, at which point it can be severed from the parent plant and transplanted.

Grafting

Grafting involves joining two plants together so that they grow as one. This method is commonly used in fruit tree propagation, allowing growers to combine the desirable traits of two different plants. For example, a hardy rootstock can be grafted with a scion from a high-yielding fruit variety. Grafting requires precision and practice, as the vascular tissues of the two plants must align for successful fusion.

Care for Young Plants After Transplanting

Once seedlings and cuttings have established roots, they require careful care and attention to ensure healthy growth. Watch the transplanted seedlings for a few weeks to check if they're facing any stress.

Proper watering is essential; young plants have delicate root systems and can be easily damaged by overwatering or underwatering. Water the plants thoroughly, but allow the soil to dry slightly between waterings to prevent root rot or fungus gnats. Also, water at the very bottom to target root systems instead of watering on upper surfaces, this also prevents fungus gnats.

Fertilization is also important during this stage, as young plants require nutrients for healthy development. Use a balanced, water-soluble fertilizer at a reduced strength to avoid overfeeding, which can burn the tender roots. Not providing the fertilizer, however, can also delay the root growth. Gradually increase the fertilizer concentration as the plants grow and mature. This fertilizer can be provided via the irrigation system or misting system.

As the young plants continue to grow, they will need to be acclimatized to outdoor conditions if they are to be transplanted outside. This process, known as hardening off, involves gradually exposing the plants to outdoor temperatures, sunlight, and wind. Start by placing the plants in a sheltered location for a few hours each day, gradually increasing the time and exposure over a period of one to two weeks. This gradual adjustment helps the plants develop the resilience needed to thrive outdoors.

Another factor to consider is not to transplant the ready-to-be-transplanted seedlings very deep. Exception to this rule can be tomatoes, but no other greenhouse plants need to be planted deep in the newer pot.

Timing and Scheduling

Timing is a crucial aspect of successful propagation. Different plants have varying ideal times for propagation, depending on their growth cycles and environmental requirements. Understanding the best times to start seeds, take cuttings, or divide plants ensures optimal growth and reduces the risk of failure. A good signal is when the plants are in fresh early growth. For example, it can be in early spring or early winter, of course, depending upon the plant type.

For seed starting, consult seed packets or gardening guides to determine the appropriate planting time. Some seeds are best started indoors several weeks before the last expected frost date, while others can be sown directly in the greenhouse. Similarly, the timing for taking cuttings or dividing plants varies based on the plant species and local climate conditions.

Creating a propagation schedule helps organize the process and ensures that each step is completed at the right time. All the schedule of transplanting and propagation depends on the type of seasonal plant you are growing and doing that in a greenhouse, can give you extended periods of growth. Keep detailed records of planting dates, germination rates, and growth observations. This information is invaluable for planning future propagation efforts and refining techniques.

Troubleshooting Propagation Issues

Despite careful planning and execution, propagation can sometimes encounter issues. Common problems include poor germination, slow rooting, and disease outbreaks. Identifying and addressing these issues promptly is essential to prevent crop losses.

If seeds fail to germinate, consider factors such as seed quality, planting depth, and environmental conditions. Adjusting these factors may improve germination rates. For slow-rooting cuttings, ensure that the cuttings are taken from healthy, disease-free parent plants and that the growing medium is consistently moist but not waterlogged.

Fungal diseases can be a significant challenge in propagation. Maintain proper sanitation practices, such as using sterile tools and containers, and avoid overcrowding plants. Fungicides can be used as a preventive measure, but it's essential to follow label instructions and safety precautions.

Be patient and persistent. Propagation is both an art and a science, and success often comes with practice and experience. Learn from each attempt, whether successful or not, and use that knowledge to improve your future propagation endeavors. With time, you'll develop the skills and confidence to propagate a diverse range of plants, contributing to the beauty and productivity of your greenhouse garden.

CHAPTER 12

Planting and Growing Techniques in a Greenhouse

"In the greenhouse of life, every plant has its season and its need for care. Cultivate with patience, and growth will follow."

Successfully planting and growing plants in a greenhouse involves more than just placing them in soil. It requires a comprehensive understanding of various techniques to ensure your plants thrive in their controlled environment.

Proper Planting Techniques

Proper planting is the foundation of a successful greenhouse garden. It starts with selecting the right plants for your greenhouse environment and then ensuring they are planted correctly to foster healthy growth.

Before planting, the preparation of the growing medium is crucial. As I mentioned in Chapter 9, using a soilless potting mix is often preferable in a greenhouse setting due to its sterility and optimal drainage properties. This type of mix helps to ensure a healthy environment for plant roots and can be particularly beneficial for achieving consistent and reliable plant growth. Unlike garden soil, which can be too dense and harbor pests or diseases, a soilless mix offers a lightweight, airy structure that supports vigorous root growth.

As I mentioned in Chapter 11, managing plant spacing is essential to prevent overcrowding, which can lead to competition for light, water, and nutrients. Adequate spacing also improves air circulation, reducing the likelihood of disease and promoting healthier plant growth.

Refer to spacing guidelines specific to each plant species; these guidelines can typically be found on seed packets or plant labels. As a general rule, leave sufficient space between plants to allow for their mature size. This not only facilitates air movement but also makes maintenance tasks such as pruning and harvesting easier.

The planting depth is another critical factor. For seedlings, make sure that they are planted at the same depth as they were in their nursery pots. Planting too deep can cause the stems to rot while planting too shallow can leave the roots exposed and prone to drying out.

For seeds, follow the depth recommendations provided, which generally suggest planting at a depth of two to three times the seed's diameter. Cover the seeds lightly with soil and gently firm it down to ensure good soil-to-seed contact.

Watering immediately after planting is essential to settle the soil around the roots and eliminate air pockets. This initial watering helps to establish a good soil-root contact, which is vital for nutrient uptake. Depending on the plant type and greenhouse conditions, you may need to adjust your watering schedule.

Mulching can be beneficial in a greenhouse setting. While it's not always necessary, a thin layer of mulch can help retain soil moisture, suppress weeds, and regulate soil temperature. Organic mulches, such as straw or compost, also break down over time, adding nutrients to the soil. However, be mindful of pests that may shelter in mulch, and use it judiciously, particularly around tender seedlings.

Watering Requirements

Watering is one of the most crucial aspects of plant care in a greenhouse, where the controlled environment requires careful management of moisture levels.

In a greenhouse, the frequency and amount of watering depend on several factors, including the type of plants, the stage of growth, the potting medium, and the climate conditions inside the greenhouse. Young seedlings, for instance, need consistent moisture to establish their root systems, while mature plants may require less frequent watering but in larger amounts.

To determine when to water, regularly check the moisture level of the soil. This can be done using a soil moisture meter or simply by feeling the soil with your finger. The goal is to keep the soil evenly moist but not soggy. For most plants, allow the top inch of soil to dry out slightly before watering again. However, this can vary; for example, plants like succulents and cacti prefer a drier soil mix and less frequent watering.

The method of watering can also make a difference. Overhead watering with a hose or watering can is common, but it's essential to do this carefully to avoid wetting the foliage, which can promote disease. Instead, water at the base of the plants, directing the water to the soil where it can be absorbed by the roots.

For a more efficient approach, consider installing a drip irrigation system. This system delivers water directly to the root zone, reducing water waste and keeping the foliage dry. Drip systems can also be automated with timers, ensuring consistent watering schedules even when you're not available.

Water quality is another important consideration. Greenhouse plants, particularly those grown in containers, can be sensitive to the mineral content of water. If your water source has high levels of salts or other minerals, it can lead to nutrient imbalances or buildup in the soil. Using rainwater or distilled water can help mitigate these issues, especially for sensitive plants.

During the hot summer months, greenhouse temperatures can rise significantly, increasing the water needs of your plants. In these conditions, consider watering early in the morning or late in the evening to minimize evaporation and reduce water stress on the plants. In contrast, during cooler periods, water is used less frequently to avoid over-saturation of the soil.

It's also crucial to adjust your watering practices based on the specific needs of different plants. For example, fruiting vegetables like tomatoes and peppers require more water when they start setting fruit, while leafy greens like lettuce need consistent moisture throughout their growth cycle.

Additionally, plants in clay pots may require more frequent watering than those in plastic pots, as clay is porous and allows water to evaporate more quickly.

Ensure that all pots and containers have adequate drainage holes, and if you're using greenhouse beds, make sure they are designed to allow excess water to drain away. Raised beds can be particularly effective in preventing water accumulation.

Finally, monitor the overall humidity levels in the greenhouse, as they can affect the water needs of your plants. High humidity can reduce the rate of water evaporation from the soil, meaning plants may need less frequent watering. Conversely, low humidity can increase the rate of water loss through transpiration, necessitating more frequent watering.

Fertilization

Fertilization supplies the essential nutrients plants need to grow, flower, and fruit. In a controlled environment like a greenhouse, where plants are frequently grown in soilless media or containers, maintaining a consistent nutrient supply becomes especially important.

The three primary nutrients that plants need are nitrogen (N), phosphorus (P), and potassium (K), often referred to as macronutrients. Nitrogen is essential for leafy growth, phosphorus supports root and flower development, and potassium is crucial for overall plant health and disease resistance.

In addition to these, plants also require secondary nutrients like calcium, magnesium, and sulfur, as well as trace elements such as iron, manganese, and zinc. While commercial fertilizers are commonly used to supply these nutrients, sustainable gardening enthusiasts may prefer natural alternatives. For instance, compost and well-rotted manure are excellent sources of nitrogen and phosphorus. Seaweed and wood ash can provide potassium, while crushed eggshells and bone meal are rich in calcium.

Selecting the right fertilizer depends on the specific needs of the plants you are growing. For general use, a balanced fertilizer with equal parts N-P-K, such as a 10-10-10 formula, can be used. However, different plants may have specific nutrient requirements. For instance, fruiting plants like tomatoes benefit from a higher phosphorus content to support flower and fruit production, while leafy greens may require more nitrogen for lush foliage.

Fertilizers come in various forms, including granular, liquid, and slow-release formulations. Granular fertilizers are applied to the soil and watered in, releasing nutrients over time. Liquid fertilizers are mixed with water and applied directly to the soil or foliage, providing a quick nutrient boost. Slow-release fertilizers are coated with a material that dissolves slowly, providing a steady supply of nutrients over several months.

For container-grown plants, liquid fertilizers are often preferred as they provide immediate nutrition and are easy to apply. However, they need to be applied more frequently, typically every two to four weeks, depending on the plant's growth stage and nutrient needs. In contrast, slow-release fertilizers can reduce the frequency of applications but may not provide the immediate boost that some plants require.

When applying fertilizer, it's essential to follow the recommended rates and avoid over-fertilization, which can lead to nutrient imbalances and harm the plants. Over-fertilization can cause salt buildup in the soil, leading to root burn and other issues. It's better to apply less fertilizer more frequently than to apply a large amount all at once.

In addition to chemical fertilizers, organic fertilizers and compost can be excellent nutrient sources. Organic options like composted manure, worm castings, and bone meal provide a slow-release source of nutrients and improve soil structure and microbial activity. In a greenhouse, where the soil is often reused, incorporating organic matter can help maintain soil fertility and health over the long term.

Fertilization needs also change throughout the plant's life cycle. For instance, during the vegetative stage, plants require more nitrogen for growth. As they enter the flowering and fruiting stages, their need for phosphorus and potassium increases. Adjusting your fertilization schedule to match these changing needs can result in healthier, more productive plants.

Another critical aspect of fertilization is monitoring and adjusting based on plant responses. Signs of nutrient deficiencies, such as yellowing leaves, stunted growth, or poor fruit development, can indicate a need for additional nutrients. Conversely, dark green leaves and excessive vegetative growth may signal an excess of nitrogen.

Water quality can affect nutrient availability. High levels of dissolved salts in the water can interfere with nutrient uptake, particularly in container-grown plants. If you notice nutrient deficiencies despite adequate fertilization, consider testing your water quality and adjusting your irrigation practices as needed.

Adequate fertilization according to your greenhouse plants' needs can significantly enhance growth, yield, and overall plant health. By carefully selecting and applying fertilizers, you can ensure your plants receive the nutrients they need to thrive in a controlled greenhouse environment.

Pest Control

Pest control in a greenhouse is essential for maintaining plant health and productivity. The warm, humid conditions that favor plant growth can also attract pests. Effective pest management involves a combination of prevention, monitoring, and control strategies to minimize damage and protect your plants.

The first line of defense in pest control is prevention. This starts with keeping a clean and tidy greenhouse environment. Regularly remove dead leaves, plant debris, and weeds that can harbor pests. Clean and disinfect benches, tools, and equipment to prevent the spread of pests and diseases. When introducing new plants, inspect and quarantine them to ensure they are free from pests and diseases before adding them to your greenhouse.

Integrated Pest Management (IPM) is a comprehensive approach that combines various strategies to manage pests effectively. IPM emphasizes using the least harmful methods first and resorting to chemical controls only when necessary. This approach includes cultural, biological, and mechanical controls.

Cultural controls involve practices that make the greenhouse environment less hospitable to pests. For example, proper plant spacing ensures good air circulation, reducing the risk of fungal diseases. Adjusting watering schedules prevents conditions that favor pests like fungus gnats. Crop rotation and selecting pest-resistant plant varieties also help reduce pest problems.

Biological controls use natural predators or parasites to manage pest populations. Introducing beneficial insects, such as ladybugs, predatory mites, or parasitic wasps, can help control common pests like aphids, spider mites, and whiteflies. These beneficial insects can be purchased and released into the greenhouse to naturally seek out and consume pests, reducing the need for chemical pesticides.

Mechanical controls involve physical methods to remove or exclude pests. This can include using sticky traps to catch flying insects, handpicking larger pests like caterpillars, or using barriers such as row covers or screens to keep pests out. Regular plant inspections allow for early detection and intervention, preventing larger infestations.

When pests become a significant issue despite these measures, chemical controls may be necessary. However, it's crucial to use pesticides judiciously and as a last resort. Choose pesticides specifically labeled for greenhouse use and that target the pests you are dealing with. Follow label instructions carefully, including dosage and application methods. Opt for organic or less-toxic pesticides, such as insecticidal soaps or neem oil, which are safer for the environment and non-target organisms.

Monitoring is a key component of any pest control strategy. Regularly check your plants for signs of pest activity, such as holes in leaves, discolored foliage, or the presence of insects. Yellow sticky traps placed around the greenhouse can help monitor flying insect populations like whiteflies and aphids. Keep detailed records of pest sightings, treatments applied, and their outcomes to adjust your pest management practices and improve their effectiveness over time.

Additionally, maintaining plant health through proper watering, fertilization, and other cultural practices can make plants more resilient and less susceptible to pest damage. Healthy plants are better able to withstand pest attacks.

Disease Prevention

Disease prevention is a critical component of greenhouse gardening, as the controlled environment can sometimes lead to conditions that favor the spread of diseases. Greenhouses provide a warm, humid climate that can be ideal for plant growth, but these same conditions can also promote the development of fungal, bacterial, and viral diseases. To protect your plants and ensure a healthy garden, it's essential to implement effective disease prevention strategies.

The first step in disease prevention is to select disease-resistant plant varieties whenever possible. Many plant breeders offer cultivars that are bred specifically for resistance to common greenhouse diseases. By choosing these varieties, you can reduce the likelihood of disease outbreaks and minimize the need for chemical treatments.

Good sanitation practices are crucial in preventing the spread of diseases. Start by thoroughly cleaning the greenhouse before each growing season. Disinfect surfaces, tools, and containers with a solution of one part bleach to nine parts water or a commercial disinfectant. Remove and dispose of any plant debris, as it can harbor pathogens that may infect new plants. Keep the greenhouse free of weeds, which can serve as reservoirs for diseases.

As discussed in previous chapters, proper spacing, pruning, and thinning is crucial for maintaining good airflow and circulation. Humidity levels, also need to be controlled by using fans, dehumidifiers, and adjusting watering practices. This prevents the growth of unwanted pathogens.

Monitoring your plants regularly for signs of disease is key to early detection and intervention. Look for symptoms such as discolored or wilting leaves, unusual spots or lesions, stunted growth, and mold or mildew on the surface of leaves or stems. If you notice any signs of disease, promptly remove and dispose of the affected plant material. Infected plants should be isolated from healthy ones to prevent the spread of pathogens.

Biological control methods can also be used to prevent and manage diseases in the greenhouse. Beneficial microorganisms, such as mycorrhizal fungi and beneficial bacteria, can be introduced to the soil to promote plant health and suppress harmful pathogens. These organisms form symbiotic relationships with plant roots, enhancing nutrient uptake and providing a natural defense against soil-borne diseases.

In cases where diseases do occur, it's important to accurately identify the pathogen and choose the appropriate treatment. Fungicides, bactericides, and other chemical treatments should be used as a last resort and applied according to label instructions. Organic and natural alternatives, such as neem oil or copper-based sprays, can be effective against certain diseases and are often less harmful to the environment.

Avoid planting the same species or related species in the same location year after year. Instead, rotate crops with different nutrient and disease profiles to reduce the buildup of pathogens in the soil.

Pruning and Maintenance

Pruning and maintenance are essential practices in greenhouse gardening, helping to ensure healthy plant growth, maximize yields, and maintain an aesthetically pleasing environment. Proper pruning techniques not only improve plant structure and appearance but also play a crucial role in disease prevention, air circulation, and light penetration. In this section, we will explore the importance of pruning, the tools needed, and the best practices for maintaining your greenhouse plants.

Pruning serves multiple purposes, including removing dead or diseased plant material, shaping plants, encouraging new growth, and increasing flower and fruit production. It is a practice that requires a good understanding of plant biology and growth habits, as different plants respond differently to pruning.

One of the primary reasons for pruning is to remove dead, damaged, or diseased parts of the plant. This not only improves the plant's appearance but also prevents the spread of diseases and pests. Dead branches or leaves can harbor pathogens, and removing them reduces the risk of infection. Similarly, pruning out damaged areas helps plants focus their energy on healthy growth rather than trying to repair damaged tissue.

Another critical aspect of pruning is thinning, which involves selectively removing branches or stems to improve air circulation and light penetration. In a greenhouse, where space can be limited and plants are often densely planted, thinning is essential to prevent overcrowding.

Good air circulation helps reduce the humidity around plants, lowering the risk of fungal diseases. It also ensures that all parts of the plant receive adequate light, which is vital for photosynthesis and overall plant health.

The timing of pruning is also important. While some plants can be pruned at any time of the year, others have specific pruning seasons. For example, flowering plants are often pruned after they have bloomed to avoid removing buds that would produce the next season's flowers.

Fruit-bearing plants, on the other hand, may require pruning during their dormant season to encourage fruiting the following year. Understanding the specific needs of each plant type in your greenhouse will help you determine the best time for pruning.

When it comes to pruning tools, having the right equipment is essential for making clean cuts that heal quickly. Common pruning tools include pruning shears, loppers, and pruning saws. Pruning shears are ideal for cutting small branches and stems, while loppers are used for thicker branches.

Pruning saws are necessary for removing large branches or making precise cuts in tight spaces. It's important to keep all pruning tools sharp and clean to prevent the spread of diseases. Sterilize tools with a solution of one part bleach to nine parts water or use rubbing alcohol, especially when moving from one plant to another.

In addition to regular pruning, maintenance tasks in the greenhouse include watering, fertilizing, and monitoring for pests and diseases. Keeping a close eye on the health of your plants allows you to address issues early before they become major problems.

For instance, if you notice yellowing leaves, it could indicate a nutrient deficiency, which can be addressed by adjusting your fertilization schedule. Similarly, spotting pests early allows you to implement control measures before they can cause significant damage.

Proper irrigation is another critical aspect of maintenance. Watering needs can vary significantly depending on the type of plants, the growing medium, and the greenhouse conditions. It's important to provide consistent moisture levels without overwatering, which can lead to root rot and other problems. Automated irrigation systems, such as drip irrigation, can help maintain consistent moisture levels and reduce the risk of overwatering.

Weeding is another maintenance task that should not be overlooked. Weeds compete with your plants for water, nutrients, and light, and can also harbor pests and diseases. Regularly remove weeds from the greenhouse, taking care to pull them out by the roots to prevent regrowth.

Regular inspections of the greenhouse structure and equipment are necessary to ensure everything is in good working order.

In summary, pruning, and maintenance are ongoing tasks in greenhouse gardening that require attention to detail and a good understanding of plant care. By keeping plants healthy and well-maintained, you can create a thriving greenhouse garden that produces abundant flowers, fruits, and vegetables. Regular pruning, proper watering, vigilant pest and disease monitoring, and consistent maintenance are the keys to success in this controlled growing environment.

CHAPTER 13
Maintaining Your Greenhouse Throughout the Year

"The success of a greenhouse is not just in its construction, but in its daily care and upkeep."

Maintaining your greenhouse throughout the year is a vital aspect of ensuring its efficiency and longevity. Just as you would take care of your home, your greenhouse requires regular attention to thrive in every season.

Whether you're preparing for the changing seasons, troubleshooting issues, or simply performing routine maintenance, keeping on top of these tasks will help create an optimal growing environment for your plants.

Seasonal Maintenance Task & Preparation

Maintaining a greenhouse isn't just about setting it up and forgetting it. It requires ongoing attention and care throughout the year. Each season introduces unique tasks and considerations to ensure your greenhouse remains in optimal condition and continues to support your plants effectively. Let's dive into the essential maintenance tasks you'll need to perform throughout the seasons.

Spring Preparation: Embrace Renewal

As winter recedes, your greenhouse will need a thorough spring cleaning. Begin by clearing out any debris or dead plant material that may have accumulated. This is also the perfect time to inspect your greenhouse structure for any damage caused by winter weather. Check the frame, glazing, and ventilation systems for issues like cracks or leaks. Clean the glass or polycarbonate panels to maximize light penetration. Ensure that the heating system is working efficiently, as spring temperatures can fluctuate.

In spring, you'll also need to plan for the new growing season. Check your soil or growing medium and amend it as needed. Prepare your planting areas by organizing and cleaning pots, trays, and other equipment. Ensure that your irrigation system is set up and ready to handle the increased watering needs of spring plants. Finally, consider any new plants or seeds you plan to introduce and prepare accordingly.

Summer Preparation: Monitor and Maintain

During the summer months, ventilation and cooling become critical. Check that your ventilation systems are functioning properly, to avoid overheating. Clean and maintain fans, louvers, and vents to ensure they operate smoothly. Adjust shading systems to protect your plants from excessive sun exposure. Regularly check the temperature and humidity levels inside the greenhouse, adjusting as needed. Also, inspect the irrigation system to prevent any water shortages that could stress your plants. Plan for any potential pest issues that often arise in summer. Implement integrated pest management strategies to keep pests at bay.

Autumn Preparation: Ready for the Cold

As temperatures drop, start preparing your greenhouse for winter. This includes checking and repairing any damage from the summer and ensuring that the heating system is ready for the cold weather. Clean and sterilize your tools and greenhouse surfaces to prevent the spread of diseases. This is also a good time to prune and tidy up any overgrown plants. Store any summer equipment that won't be used over the winter, such as shade cloths.

Consider adjusting your planting schedule to account for the shorter daylight hours and cooler temperatures. If you have any late-season crops, ensure they are properly protected from potential frost. Keep an eye on the weather forecast and be prepared for any early cold snaps.

Winter Preparation: Protect and Preserve

The season of vigilance. Winter maintenance involves monitoring for issues related to cold weather. Ensure that your heating system is functioning correctly and efficiently. Inspect the greenhouse for any issues with insulation or drafts. Keep an eye on snow accumulation on the roof, as excessive snow can cause structural damage. Regularly check the greenhouse's temperature and humidity levels to ensure that your plants remain protected from frost and extreme cold. Ensure that your irrigation system is functioning correctly, as water needs may change with the colder temperatures. Keep an eye on snow accumulation on the roof and remove it as needed.

Throughout the year, keep a detailed maintenance log to track repairs, replacements, and observations. This will help you stay organized and address recurring issues before they become major problems. Regular maintenance ensures that your greenhouse remains a thriving environment for your plants, regardless of the season.

Troubleshooting Common Issues

Even with the best planning and maintenance, you may encounter issues with your greenhouse. Knowing how to troubleshoot common problems can help you address them effectively and minimize their impact on your plants. Let's explore some typical greenhouse issues and how to resolve them.

Ventilation Problems: If you notice that your greenhouse is becoming too hot or humid, check your ventilation systems for blockages or malfunctions. Ensure that vents, fans, and louvers are clean and operating correctly. Consider adjusting the ventilation settings to improve air circulation and temperature control.

Temperature Fluctuations: If you experience significant temperature swings, check your heating and cooling systems for efficiency. Ensure that your insulation is adequate and that there are no drafts or leaks in the greenhouse structure. Use thermostats and climate control systems to maintain a stable temperature.

Pest Infestations: Pests can quickly become a problem in a greenhouse environment. Regularly inspect your plants for signs of pests such as aphids, spider mites, or whiteflies. Implement integrated pest management strategies, such as using beneficial insects, applying organic pesticides, or introducing physical barriers. Keep your greenhouse clean and free of plant debris to reduce pest habitats.

Disease Outbreaks: Diseases such as mildew, mold, or bacterial infections can spread rapidly in a greenhouse. To prevent disease outbreaks, practice good sanitation by cleaning and sterilizing your greenhouse regularly. Remove any diseased plants and dispose of them properly. Monitor your plants for symptoms of disease and take action promptly to address any issues.

Irrigation Issues: Irregular or insufficient watering can affect plant health. If you notice that your plants are not receiving adequate water, check your irrigation system for clogs or malfunctions. Adjust your watering schedule based on the needs of your plants and the current climate conditions. Ensure that your irrigation system provides consistent and even water distribution.

Structural Problems: Over time, greenhouse structures can develop issues such as leaks, cracks, or damage. Regularly inspect the frame, glazing, and other components for signs of wear or damage. Repair any issues promptly to prevent further deterioration. Ensure that your greenhouse is structurally sound and capable of withstanding environmental stresses.

Regular Cleaning and Upkeep

Maintaining a clean and organized greenhouse is essential for the health of your plants and the efficiency of your greenhouse operations. Regular cleaning and upkeep help prevent diseases, pests, and other issues that can affect plant growth. Let's delve into the key aspects of greenhouse cleaning and maintenance.

Daily Tasks: Start by performing daily tasks to keep your greenhouse in top condition. This includes checking for any signs of pests or diseases and addressing them immediately. Water your plants as needed and inspect the irrigation system to ensure it is functioning properly. Remove any dead or wilted plant material to prevent the spread of disease.

Weekly Tasks: On a weekly basis, perform more thorough inspections and cleaning. Check the temperature and humidity levels and adjust as necessary. Clean the greenhouse surfaces, including the floors and shelves, to remove any accumulated dirt or debris. Inspect the ventilation and heating systems for any issues, and clean the vents and fans to ensure proper airflow.

Monthly Tasks: Monthly tasks involve deeper cleaning and maintenance. Clean the glass or polycarbonate panels to maximize light penetration. Inspect the greenhouse structure for any damage or wear and address any issues promptly. Sterilize your tools and equipment to prevent the spread of diseases. Check and maintain your greenhouse's heating and cooling systems, ensuring they are functioning efficiently.

Seasonal Tasks: Seasonal cleaning tasks are crucial for preparing your greenhouse for the changing weather. In spring, perform a thorough cleaning to remove winter debris and prepare for the growing season. In summer, focus on maintaining ventilation and cooling systems. In autumn, clean and sterilize the greenhouse to prevent disease and prepare for winter. In winter, monitor and address any issues related to cold weather, such as insulation and heating.

Organizational Tips: Keeping your greenhouse organized helps streamline maintenance and operations. Store tools and equipment in designated areas to prevent clutter and ensure they are easily accessible. Label containers and planting areas to keep track of your plants and their needs. Regularly review and update your maintenance log to track tasks and observations.

By staying on top of these tasks, you can ensure that your plants thrive and your greenhouse remains a successful growing space.

Equipment Maintenance

Equipment maintenance is a critical aspect of greenhouse management. Properly maintaining your greenhouse equipment ensures that it operates efficiently and effectively, reducing the risk of breakdowns and prolonging its lifespan. Here's a comprehensive guide to keeping your greenhouse equipment in top shape.

Heating Systems: Regular maintenance of your heating system is essential for maintaining a stable environment for your plants. Check the system's filters and replace them as needed to ensure efficient operation. Clean the heating elements and inspect the thermostat to ensure accurate temperature control. Test the system before the cold weather sets in to address any issues promptly.

Cooling Systems: Cooling systems, such as fans and evaporative coolers, require regular maintenance to ensure they function effectively. Clean the fans and cooling pads to remove any dust or debris. Check the belts and bearings for wear and replace them if necessary. Inspect the cooling system's water supply and ensure it is clean and functioning properly.

Ventilation Systems: Proper maintenance of ventilation systems is crucial for managing temperature and humidity levels. Clean and inspect vents, louvres, and fans to ensure they operate smoothly. Check for any obstructions that may impede airflow and address them promptly. Test the ventilation system to ensure it responds correctly to temperature and humidity changes.

Irrigation Systems: Regular maintenance of your irrigation system is essential for providing consistent and adequate water to your plants. Check the system's filters and flush them to remove any clogs. Inspect the hoses, nozzles, and emitters for leaks or damage and repair or replace them as needed. Test the system to ensure even water distribution.

Tools and Equipment: Clean and maintain your gardening tools and equipment regularly to ensure they remain in good condition. Sharpen blades, lubricate moving parts, and repair any damage.

Store tools in a dry and organized manner to prevent rust and wear. Regularly inspect and replace any worn or damaged equipment.

Safety Equipment: Ensure that your safety equipment, such as gloves, goggles, and masks, is in good condition. Replace any damaged or worn safety gear to protect yourself while working in the greenhouse. Regularly inspect safety equipment and keep it readily accessible.

CHAPTER 14
Popular Greenhouse Plants

"The greenhouse is used for flowering and vegetative plants as well as ornamental plants such as different types of houseplants"

Greenhouses are used to grow different types of plants, such as vegetative, fruiting, and house plants. They provide an ideal environment for growing this variety of plants by offering control over factors like temperature, humidity, and light. Understanding the specific needs of each plant type can help you optimize your greenhouse setup and achieve a successful harvest.

Vegetative and Fruiting Plants

The different types of vegetative and fruiting plants grown in greenhouses are discussed below with their major requirements. Requirements vary depending upon the weather conditions but greenhouses can be used to aid the process of seedling growth and then later on when ideal conditions reach, they can be transplanted to gardens, as discussed in chapter 11.

Tomato

- Sunlight: 6-8 hours of direct sunlight per day.
- Water: Consistent and deep watering at the base of the plant.
- Soil: Well-drained, fertile soil with pH 6.2-6.8.
- Temperature: The optimal temperature required during the day is 70-85°F (21-29°C) and 55-70°F (13-21°C) at night.
- Humidity: Around 60-70%.
- Nutrients: Balanced fertilizer early; higher phosphorus and potassium during fruiting.
- Airflow: Good circulation; space plants 18-24 inches apart.

- Space: 18-24 inches for determinate varieties, 24-36 inches for indeterminate varieties.
- Pests and Diseases: Regular monitoring; organic or chemical treatments as needed.
- Pruning and Training: Remove suckers; use stakes or cages for support.
- Best Period to Start: Start seeds 6-8 weeks before the last frost date.
- Ready to Harvest: 50-80 days from transplanting.
- Germination Condition: Seeds require 70-80°F (21-27°C) and consistent moisture.

Cucumber

- Sunlight: 6-8 hours of direct sunlight per day.
- Water: Regular watering, keeping the soil consistently moist but not waterlogged.
- Soil: Well-drained, rich soil with pH 6.0-7.0.

- Temperature: The optimal temperature required during the day is 70-85°F (21-29°C) and 55-65°F (13-18°C) at night.
- Humidity: High; 70-80%.
- Nutrients: Balanced fertilizer; higher potassium during fruiting.
- Airflow: Ensure good ventilation; space plants 12-18 inches apart.
- Space: Allow ample room for vines to spread.
- Pests and Diseases: Watch for aphids and mildew; use appropriate treatments.
- Pruning and Training: Train vines on a trellis to save space and improve airflow.
- Best Period to Start: Start seeds 4-6 weeks before planting.
- Ready to Harvest: 50-70 days from planting.
- Germination Condition: Seeds need 70-80°F (21-27°C) and consistent moisture.

Pepper

- Sunlight: 6-8 hours of direct sunlight per day.
- Water: Regular, consistent watering; keep soil evenly moist.
- Soil: Well-drained soil with pH 6.0-6.8.
- Temperature: The optimal temperature required during the day is 70-85°F (21-29°C) and 55-70°F (13-21°C) at night.
- Humidity: 50-70%.
- Nutrients: Balanced fertilizer; higher potassium during fruiting.
- Airflow: Ensure good circulation; space plants 12-18 inches apart.
- Space: 12-18 inches between plants.
- Pests and Diseases: Monitor for aphids and spider mites; use suitable treatments.

- Pruning and Training: Remove lower leaves; support plants if needed.
- Best Period to Start: Start seeds 8-10 weeks before transplanting.
- Ready to Harvest: 60-90 days from transplanting.
- Germination Condition: Seeds require 70-85°F (21-29°C) and consistent moisture.

Lettuce

- Sunlight: 4-6 hours of direct sunlight per day.
- Water: Regular, consistent watering; keep the soil moist.
- Soil: Well-drained, fertile soil with pH 6.0-7.0.
- Temperature: The optimal requirement during the day is 60-70°F (15-21°C) and 50-60°F (10-15°C) at night.

- Germination conditions: Seeds need 60-70°F (15-21°C) and consistent moisture.
- Humidity: Moderate; around 60-70%.
- Nutrients: Balanced fertilizer.
- Airflow: Good circulation; space plants 6-12 inches apart.
- Pests and diseases: Watch for aphids and mildew; use appropriate treatments.
- Pruning and Training: Not typically needed; harvest outer leaves as needed.
- The best period to start: Start seeds 6-8 weeks before transplanting.
- Harvesting: 30-60 days from planting.

Basil

- Sunlight: 6-8 hours of direct sunlight per day
- Water: After every 4-7 days depending upon the climate.
- Soil: Fertile soil with a pH of 6.0-7.0
- Temperature: 24-29°C (80-90°F) is the best range for growing basil.
- Germination conditions: 70-80°F (21-25 °C) with appropriate moisture that doesn't dry the soil or make it 'too' wet.
- Humidity: 40-60%
- Airflow: Good circulation is needed; space plants 12 to 16 inches apart.
- Pests and diseases: ray mold, Cercospora leaf spot, downy mildew, root rot, snail infestation, aphids.
- Pruning and training: better to prune in the early growing season. Do it every 3-4 weeks.
- The best time to start is in the late spring or early summer, or when temperatures begin to rise above 10 degrees Celsius.
- Harvesting: Best when the plant reaches 8-10 inches in height, or approximately 60-70 days after planting.

Cilantro

- Sunlight: 6 hours of direct sunlight per day.
- Water: one inch of water every week.
- Soil: fertile soil with good drainage required for cilantro. pH must be between 6.2-6.8
- Temperature: 50-85°F optimum for growth.
- Humidity: Average 50% humidity is ideal for cilantro.
- Germination requirements: 50-70°F (10-21°C) with moderate, consistent moisture.
- Spacing and airflow: Seedlings need to be placed 2-4 inches apart at the transplanting stage.

- Pests and diseases: Look out for aphids, nematodes, and weeds. Some common diseases include *Alternaria* leaf spot, cilantro yellow blotch, bacterial leaf spot, damping-off, and powdery mildew.
- The best period to start:
- Pruning and harvesting: When stems reach 6 inches, cut them just above the ground. Keep cutting any large leaves, and if leaves start turning yellow.

Mint

- Sunlight: Mint grows well in sunlight as well as shade. 6 to 8 hours of sunlight is good.
- Watering: It is done whenever the inch of top soil layer starts to feel dry. Frequent watering is better for young plants.
- Soil: moist, well-drained, and fertile soil with a pH of 6 to 7 is best for mint.

- Temperature: 50-85°F is an ideal range but temperatures above 100 °F (37°C) can affect growth
- Humidity: 40-50%
- Germination requirements: 55-75°F and 60-70% of humidity
- Pests and diseases: Look out for aphids, leafhoppers, and mites. Common diseases are mint rust and verticillium wilt.
- The best period to start: It is in the late spring or early summer. The best is to plant 4-6 weeks before the last frost.
- Pruning and harvesting: It is done when the stems reach 3-4 inches in length.

Rosemary

- Sunlight: It requires at least 6 hours of sunlight
- Watering: Every 1-2 weeks or it is done whenever the inch of the top soil layer starts to dry.
- Soil: fertile, loamy soil is best for rosemary, with a pH between 6 to 7.
- Temperature: 68-86°F (20-30°C) during late spring and early summer. Temperatures above 15°C (59°F) are ideal for growth of rosemary.
- Humidity: 40-60%
- Germination requirements: 18-24°C of soil temperature and pre-moistened soil with warm water. Keep the surface of the soil consistently moist but not soaking wet.
- Pests and diseases: leafhoppers, rosemary beetles, powdery mildew, downy mildew, root rot.
- Pruning: The best time to prune is in late spring when flowering ends.
- The best time to start rosemary: It is after the end of frost in early spring.

House Plants

Greenhouses are also used for growing ornamental plants, such as houseplants, as the temperature conditions are ideal if maintained. They also improve the aesthetic appeal of the greenhouse when grown along with seasonal vegetative plants. Below are some of the commonly used houseplants in greenhouses.

Begonias

- Sunlight: 4-6 hours of morning or late afternoon sunlight.
- Watering: Thorough watering every 2-4 days indoors when the surface of the soil feels dry.
- Soil: Soilless pots with compost and perlite are ideal to grow indoors. pH should be between 5.5-6.5.
- Temperature: 60-80°F (16-27°C), with slightly lower temperatures at night (55°F).
- Humidity: 60-80%
- Germination requirements: 60-80°F (16-27°C). A little bit variable depending upon winter and

summer maintenance in the greenhouse. Keep the pot or container consistently moist.

- Pests and diseases: Look out for spider mites, snails, nematodes, and whiteflies, as well as diseases such as bacterial leaf spot, root rot, and bacterial blight.
- Pruning: Continuous pruning throughout the year keeps the plant in shape. During the flowering season, prune dead flower stems, and after the flowering season, remove larger portions of stems.
- Best time to start: Plant begonias when frost is completely gone and nights are warm, i.e., in early spring.

Orchids

- Sunlight: 6-8 hours of indirect sunlight or 3 hours of direct sunlight daily.
- Watering: Once every week.
- Soil: They grow best in airy special mixes including sphagnum mosses (as organic material), bark, perlite, and sand.
- Temperature: 70°- 85°F (20°- 30°C) during day time. They need slightly lower temperatures during nighttime, 60°F (16°C).
- Humidity: 40-70%
- Germination requirements:

- Pests and diseases: Look out for aphids, spider mites, whiteflies, caterpillars, and mealybugs. Black rot, gray leaf spot, and mosaic virus are some of the common diseases of orchids.
- Pruning and training: Prune orchids near late fall or early winter once the flowering season ends. Cut the stem so that only an inch or so of the stem is left.
- Best time to start: It is in late spring or early summer when days are more humid.

Geraniums

- Sunlight: 6-8 hours of direct sunlight.
- Watering: It is done whenever the inch of top soil layer starts to feel dry. Let the soil dry a bit between waterings.

- Soil: Fertile and well-drained soil with a pH of 6.5.
- Temperature: 70-75°F (21-24°C) at daytime and 60-65°F (15-18°C) at nighttime.
- Humidity: 40-60%.
- Germination requirements: Best germination at 75°F with consistent moisture.
- Pests and diseases: Look out for aphids. Botrytis blight, Alternaria leaf spot, bacterial blight, Pseudomonas cichorii, Pythium root rot, and verticillium are some common diseases of geraniums.
- Best time to start: It is after the end of frost in early spring.

Fuchsias

- Sunlight: 6 hours of sunlight, especially morning sunlight.

- Watering: 1-1.5 inches of water every week. On hot days, water daily.
- Soil: Fertile, well-drained soil rich in nutrients with a pH of 6-7.
- Temperature: 55-80°F is the best range for fuchsias to grow and bloom.
- Germination requirements: 65-75°F with consistent moisture.
- Pests and diseases: Look out for thrips, armored scales, grapevine moths, and Fuchsia gall mites. Armillaria root rot, Botrytis blight, damping-off (pythium root rot), and verticillium are some common diseases of fuchsias.
- Best time to start: It is in late spring or early summer.

Ferns

- Sunlight: Ferns require indirect, bright light during the day but they need to be put in shades when temperatures rise above 75°F (or 24°C).
- Watering: Water every two weeks, but water after 4 to 7 days on hot days.
- Soil: Fertile, regular moist soil is good for indoor ferns.
- Temperature: During the daytime, temperatures should be between 65-75°F, and during nighttime, temperatures should be kept below 60°F.
- Humidity: 40-50%

- Germination requirements: 25°C (77°F) with a humid environment and moist soil.
- Pests and diseases: Look out for scales, aphids, and mealybugs.

PART 5: ADVANCED GREENHOUSE GARDENING TECHNIQUES

CHAPTER 15

Hydroponics for Beginners

"In the world of gardening, hydroponics is like an innovative twist to a classic recipe, offering a fresh take on growing plants without soil."

Hydroponics might seem like a complex science at first, but trust me, it's an incredibly rewarding way to grow plants—especially in a greenhouse setting. Unlike traditional gardening, hydroponics allows you to bypass soil altogether and grow your plants in nutrient-rich water.

This method not only optimizes space but also gives you precise control over the growing conditions, leading to faster growth and higher yields.

Introduction to Hydroponics

Hydroponics is a fascinating and increasingly popular method of gardening that allows you to grow plants without the use of soil. Instead, plants are grown in nutrient-rich water solutions, which provide all the necessary minerals and nutrients directly to the roots.

This soil-less cultivation method is particularly well-suited for greenhouse gardening, as it offers precise control over growing conditions and can yield impressive results.

Hydroponics might seem daunting at first, but it's actually a highly efficient and rewarding way to grow plants. It offers numerous benefits, including faster growth rates, higher yields, and the ability to grow in limited space. Whether you're a seasoned gardener looking to try something new or a beginner eager to explore innovative gardening techniques, hydroponics provides a unique and exciting opportunity.

Hydroponics is not a new concept. Historical evidence suggests that the Hanging Gardens of Babylon may have utilized a form of hydroponics, and ancient Aztec cultures developed floating gardens called "chinampas."

In modern times, hydroponics has become a cutting-edge technology, driven by advancements in agricultural science and a growing interest in sustainable and efficient food production.

Types of Hydroponic Systems

There are several types of hydroponic systems, each with its own advantages and best uses. Understanding these systems will help you choose the one that best suits your needs and gardening goals.

Nutrient Film Technique (NFT): This system involves a thin film of nutrient solution flowing over the roots of plants. The roots are suspended in a channel, allowing them to access the nutrients while being exposed to air for oxygen. NFT systems are popular for growing small, fast-growing plants like leafy greens and herbs. They are relatively simple to set up and maintain, making them ideal for beginners. The key to success with NFT systems is ensuring a constant flow of nutrients and maintaining a proper slope for drainage.

Deep Water Culture (DWC): DWC is one of the simplest hydroponic systems, where plants are suspended in a nutrient-rich solution. The roots are submerged in the solution, while an air pump provides oxygen to the roots. This system is great for growing larger plants, such as tomatoes and peppers, as it provides ample nutrients and oxygen. The main challenge with DWC is maintaining oxygen levels in the water, especially in warm climates, as warm water holds less oxygen.

Ebb and Flow (Flood and Drain): In this system, plants are grown in a tray filled with a growing medium, such as clay pellets or gravel. The tray is periodically flooded with nutrient solution, which then drains back into a reservoir. This cycle repeats, providing the roots with nutrients and oxygen. Ebb and flow systems are versatile and can accommodate a wide range of plants. They require careful monitoring of the flooding cycle and regular maintenance of the pump and drainage system.

Wick System: The wick system is one of the simplest hydroponic setups. It uses a wick, typically made of absorbent material, to draw nutrient solution from a reservoir into the growing medium. The plants absorb the nutrients through the medium. Wick systems are passive, meaning they don't require pumps or electricity. They are ideal for small plants and herbs but may not provide enough nutrients for larger or more demanding plants.

Drip System: In a drip system, a timer-controlled pump delivers nutrient solution to the base of each plant through a network of tubes and drip emitters. The excess solution drains back into the reservoir. Drip systems are highly versatile and can support a wide variety of plants. They allow precise control over nutrient delivery, making them suitable for both beginners and experienced gardeners. However, they require regular monitoring to prevent clogging of the drip emitters.

Aeroponics: Aeroponics is a high-tech system where plants are suspended in the air, and the roots are misted with a nutrient solution. This method provides excellent oxygenation to the roots, leading to rapid growth. Aeroponics is suitable for a variety of plants but requires precise control of the misting system and nutrient levels. It's an advanced system that may require more technical knowledge and maintenance.

Choosing the right hydroponic system depends on several factors, including the types of plants you want to grow, the available space, your budget, and your level of experience. Each system has its own set of challenges and advantages, so consider your specific needs and resources when making your decision.

Maintaining Nutrient Solutions

In hydroponics, the nutrient solution is the lifeline of your plants. It provides essential minerals and nutrients that plants would typically absorb from soil. Properly maintaining this solution is crucial for the health and growth of your plants. Let's explore how to mix, monitor, and adjust your nutrient solutions for optimal results.

Mixing Nutrient Solutions

Start with high-quality water, ideally filtered or distilled, as impurities in tap water can affect nutrient uptake. Nutrient solutions are typically made from concentrated nutrient salts that you mix with water according to specific ratios. These solutions contain a balanced blend of macro-nutrients (nitrogen, phosphorus, potassium) and micro-nutrients (calcium, magnesium, iron, etc.) that are essential for plant growth.

Follow the manufacturer's instructions carefully when mixing nutrients. Over-fertilizing can lead to nutrient burn, while under-fertilizing can result in nutrient deficiencies. It's important to measure the electrical conductivity (EC) of the solution, which indicates the concentration of nutrients. EC meters are commonly used to ensure the solution is at the appropriate strength for your plants' growth stage.

Monitoring pH Levels

The pH of the nutrient solution affects the availability of nutrients to the plants. Most hydroponic plants thrive in a pH range of 5.5 to 6.5. Outside this range, certain nutrients may become unavailable or toxic. Use a reliable pH meter to regularly check the pH of your solution. If the pH is too high, add a pH down solution (usually phosphoric acid). If it's too low, use a pH up solution (usually potassium hydroxide).

Maintaining Nutrient Balance

As plants absorb nutrients, the composition of the nutrient solution changes. It's important to regularly replace the nutrient solution to maintain a balanced supply of nutrients. Depending on the system and plant type, this can be done weekly or bi-weekly. Replacing the solution also helps prevent the buildup of harmful salts and pathogens.

Nutrient Solution Temperature

The temperature of the nutrient solution can affect plant health. Ideally, keep the solution between 65-75°F (18-

24°C). Warmer temperatures can reduce the oxygen levels in the water, while colder temperatures can slow down nutrient uptake. Use a water heater or chiller if necessary to maintain optimal temperatures.

Oxygenation

In systems where roots are submerged in water, oxygenation is crucial. Use an air pump and air stones to keep the water well-aerated. This prevents root rot and ensures that the roots receive enough oxygen. Check the air pump regularly to ensure it's working properly.

Setting Up a Hydroponic System

Setting up a hydroponic system involves several key steps, from choosing the right location to assembling the equipment and planting your first crops. Whether you're setting up a small system for personal use or a larger setup for commercial production, the principles are the same.

Choosing a Location: Select a location with ample light, preferably in a greenhouse or near large windows if you're growing indoors. Hydroponic systems require a consistent light source, whether from natural sunlight or artificial grow lights. Ensure that the area has access to electricity and water. It's also important to consider ventilation and temperature control to prevent overheating and ensure adequate air circulation.

Assembling the System: Depending on the type of hydroponic system you've chosen, the assembly process will vary. Begin by setting up the structure that will support your plants. This could be a series of channels for an NFT system, a reservoir and net pots for DWC, or trays for an ebb and flow system. Follow the manufacturer's instructions carefully to ensure everything is set up correctly.

Next, install the necessary components for nutrient delivery and oxygenation. This may include pumps, tubing, air stones, and drip emitters. If you're using a system that requires a growing medium, such as clay pellets or rock wool, prepare the medium and place it in the appropriate containers.

Lighting and Environmental Controls: If you're growing indoors or in a controlled environment, set up your lighting system. LED grow lights are ideal for hydroponic systems. Position the lights at the correct height above the plants and adjust as the plants grow. Install timers to automate the light cycles.

For temperature and humidity control, consider installing fans, heaters, and humidifiers as needed. These controls are crucial for maintaining a stable environment, especially in a greenhouse setting where temperatures can fluctuate.

Preparing the Nutrient Solution: Fill the reservoir with the solution, ensuring it's at the correct pH and EC levels. Test the system to make sure the nutrient solution is circulating properly and that there are no leaks.

Planting: Once the system is set up and running smoothly, it's time to plant your seeds or seedlings. If you're starting from seeds, use a propagation tray and transplant the seedlings into the hydroponic system once they've developed a few true leaves. If you're using seedlings, rinse the soil off the roots carefully before placing them in the system.

Monitoring and Adjusting: After the initial setup, it's crucial to monitor the system closely. Check the pH and EC levels regularly and adjust the nutrient solution as needed. Monitor the health of your plants and watch for any signs of nutrient deficiencies, pests, or diseases. Make any necessary adjustments to the environment, such as adjusting the lights or improving air circulation.

Setting up a hydroponic system requires careful planning and attention to detail. However, once established, it can provide a highly efficient and rewarding way to grow a variety of plants. By following these steps and maintaining a regular schedule of monitoring and maintenance, you can ensure a successful and productive hydroponic garden.

Choosing Hydroponic Plants

Selecting the right plants for your hydroponic system is crucial for a successful gardening experience. Not all plants thrive in hydroponic conditions, so it's important to choose species that are well-suited to this growing method. Here's a guide to selecting the best plants for your hydroponic garden.

Leafy Greens: Leafy greens are among the easiest and most popular plants to grow hydroponically. They have relatively short growing cycles and can be harvested multiple times. Varieties such as lettuce, spinach, kale, and Swiss chard thrive in hydroponic systems. They don't require extensive root systems, making them ideal for systems like NFT and DWC.

Herbs: Many herbs grow exceptionally well in hydroponic systems. Basil, mint, cilantro, parsley, and dill are popular choices. Herbs often have a high market value and can be harvested continuously, making them a great option for home gardeners and commercial growers alike. They require minimal space and can be grown in small systems.

Fruiting Vegetables: While more challenging than leafy greens and herbs, fruiting vegetables can also be grown hydroponically with the right care. Tomatoes, cucumbers, peppers, and strawberries are common choices. These plants require more space, light, and nutrients than leafy greens, so they're best suited for larger systems with strong support structures and adequate lighting.

Root Vegetables: Although less common, certain root vegetables can be grown hydroponically, particularly in systems like DWC or aeroponics. Radishes, carrots, and beets are examples. Growing root vegetables hydroponically can be challenging due to their need for space to develop underground parts, so they require careful planning and system setup.

Vining Plants: Vining plants such as beans and peas can be successfully grown in hydroponic systems. These plants require trellising or other support structures to grow vertically. They benefit from the controlled environment of a greenhouse, which protects them from pests and diseases.

Flowers and Ornamentals: Hydroponics isn't limited to edible plants. Many ornamental flowers, such as roses, orchids, and marigolds, can also be grown hydroponically. Flowers require specific nutrient mixes and light conditions, so they may need specialized care.

When selecting plants for your hydroponic system, consider factors such as growth rate, space requirements, and light needs. It's also important to think about your own preferences and goals—whether you're growing for personal consumption, aesthetic enjoyment, or commercial purposes. Start with easier crops and gradually experiment with more challenging plants as you gain experience.

Choosing the right plants is a critical step in hydroponic gardening. By selecting species that are well-suited to hydroponics and paying attention to their specific needs, you can create a productive and diverse garden that thrives year-round.

Hydroponic System Maintenance

Maintaining a hydroponic system is essential for ensuring healthy plant growth and preventing issues such as nutrient deficiencies, diseases, and equipment malfunctions. Regular maintenance tasks vary depending on the type of system you're using but generally include monitoring and adjusting nutrient solutions, cleaning components, and inspecting plants for health.

Regular Monitoring: One of the most important aspects of hydroponic maintenance is regular monitoring of nutrient solution levels, pH, and EC. Check the nutrient solution daily and top it up as needed. It's important to replace the solution regularly, typically every two weeks, to maintain a balanced nutrient profile. Use a pH meter to ensure the solution stays within the optimal range for your plants. Similarly, monitor the EC to ensure the nutrient concentration is appropriate. Regularly inspect your plants for signs of nutrient deficiencies or imbalances, such as yellowing leaves or stunted growth.

System Cleaning: Keeping your hydroponic system clean is crucial for preventing the buildup of algae, bacteria, and other contaminants. Regularly clean the reservoir, channels, and other components with a mild disinfectant. Rinse thoroughly to remove any chemical residues. If you're using a growing medium, clean or replace it periodically to prevent the buildup of salts and pathogens.

Pump and Air Stone Maintenance: Pumps and air stones are vital components in many hydroponic systems, providing circulation and oxygenation. Check pumps regularly to ensure they're functioning properly and replace any worn-out parts. Clean air stones to prevent clogging and ensure efficient oxygenation of the nutrient solution.

Pest and Disease Management: While hydroponics reduces some pest and disease risks, it's not immune. Regularly inspect plants for signs of pests or diseases, such as discolored leaves, spots, or unusual growth patterns. Implement integrated pest management (IPM) strategies, such as introducing beneficial insects, using organic pesticides, or removing affected plants.

Lighting and Environmental Controls: If you're using artificial lighting, periodically check and adjust the light levels and positions as plants grow. Ensure that timers are set correctly for the light cycle. Similarly, monitor temperature and humidity levels in the greenhouse and make necessary adjustments using fans, heaters, or humidifiers.

Record Keeping: Keeping detailed records of your hydroponic system's operation is invaluable. Record the pH, EC, nutrient concentrations, and any adjustments made. Note any issues that arise, such as pest infestations or equipment failures, and how they were resolved. This information can help you fine-tune your system and prevent future problems.

Maintaining a hydroponic system requires diligence and attention to detail, but the rewards are well worth the effort.

CHAPTER 16

Aquaponics: Integrating Fish and Plants

"In the harmony of nature, aquaponics is the perfect duet between fish and plants, creating a self-sustaining ecosystem that thrives on balance and care."

Aquaponics is an innovative and sustainable method of gardening that combines aquaculture (raising fish) with hydroponics (cultivating plants without soil). It might sound like a high-tech science project, but it's actually a beautifully simple system that anyone can set up in their greenhouse.

The basic principle is that the waste produced by the fish provides an organic nutrient source for the plants, while the plants naturally filter and clean the water, creating a balanced ecosystem. This symbiotic relationship not only conserves water but also allows you to grow fresh vegetables and raise fish simultaneously.

Here, I'll guide you through the essentials of setting up your own aquaponic system, choosing the right fish and plants, and maintaining the delicate balance needed for both to flourish.

Concept of Aquaponics

Aquaponics is a revolutionary gardening method that merges the principles of aquaculture and hydroponics to create a mutually beneficial ecosystem. At its core, aquaponics leverages the natural relationship between fish and plants, transforming waste into sustenance. In an aquaponic system, fish are kept in tanks where they produce waste. This waste, rich in nutrients, becomes the perfect fertilizer for plants. In turn, the plants absorb these nutrients, filtering and purifying the water, which is then cycled back to the fish tanks, creating a continuous loop of life-sustaining resources.

This method not only conserves water but also provides an efficient way to produce food. It's an ideal system for anyone interested in sustainable agriculture, urban farming, or maximizing their greenhouse's potential. The beauty of aquaponics lies in its simplicity and the way it mimics natural ecosystems. By understanding and harnessing these natural processes, you can create a thriving, low-maintenance garden that produces both plants and fish.

Aquaponics can be as simple or as complex as you want it to be. Beginners might start with a basic system, perhaps a small tank with a few fish and a simple grow bed for plants. More experienced gardeners might invest in larger, more elaborate setups with multiple tanks, advanced filtration systems, and automated controls. Regardless of the size or complexity of your system, the principles remain the same: fish waste provides nutrients for plants, and plants clean the water for fish.

Starting an aquaponic system requires some initial planning and setup, but once it's established, it can be incredibly rewarding. You'll need to consider the types of fish and plants you want to raise, the size and location of your system, and the necessary equipment. It's also important to understand the basics of water chemistry and how to maintain a balanced ecosystem. But don't worry—I'll guide you through each step of the process, providing tips and insights from my experience.

The concept of aquaponics might seem daunting at first, but it's actually quite intuitive once you get the hang of it. By observing and responding to the needs of your fish and plants, you'll develop a deeper understanding of how ecosystems work and how to create a harmonious environment in your greenhouse. Aquaponics not only offers a practical solution for sustainable food production but also provides a unique and engaging way to connect with nature and learn about ecological balance.

Setting Up an Aquaponic System

Setting up an aquaponic system involves several key components: a fish tank, a grow bed for plants, a water pump, and a filtration system. The first step is to choose the right location for your system. Ideally, this should be in a greenhouse or a sheltered area with plenty of natural light. If you don't have a greenhouse, you can set up your system indoors with the help of grow lights.

Next, you'll need to select a tank for your fish. This can be anything from a simple plastic tub to a more sophisticated aquarium setup. The size of your tank will depend on the number and type of fish you plan to raise. Common choices for aquaponics include tilapia, goldfish, and koi, as they are hardy and adapt well to a variety of conditions.

Once you have your tank, you'll need to set up the grow bed for your plants. This should be positioned above the fish tank to allow gravity to help with water circulation. The grow bed can be filled with a growing medium such as gravel, clay pebbles, or even coconut coir. The choice of medium will depend on the types of plants you want to grow and the specific requirements of your system.

The next step is to install a water pump to circulate water from the fish tank to the grow bed. The pump should be powerful enough to move the water efficiently but not so strong that it disturbs the fish or the plants. You'll also need a filtration system to remove solid waste and excess nutrients from the water. This can be a simple mechanical filter or a more advanced biological filter that uses beneficial bacteria to break down waste.

Once everything is set up, you can add your fish and plants. It's important to monitor the water quality regularly, checking for levels of ammonia, nitrites, and nitrates. You'll also need to keep an eye on pH levels and temperature to ensure a healthy environment for both fish and plants.

Maintaining Water Quality

Maintaining water quality is crucial in an aquaponic system. The health of your fish and the productivity of your plants depends on a delicate balance of nutrients, pH levels, and temperature. Regular testing and monitoring are essential to keep your system running smoothly.

The first thing to check is the ammonia level in your tank. Fish produce ammonia as a waste product, and high levels of it can be toxic. Beneficial bacteria in your system will convert ammonia into nitrites and then into nitrates, which are less harmful and can be used by plants as nutrients. However, if ammonia levels get too high, it can stress or even kill your fish. Regular testing and water changes can help keep ammonia levels in check.

Next, you'll want to monitor the pH levels in your system. Most fish and plants thrive in a pH range of 6.8 to 7.2, but this can vary depending on the species you are raising. If the pH gets too high or too low, it can affect nutrient uptake and the health of your fish. You can adjust pH levels using natural products like crushed coral or peat moss or with commercial pH adjusters.

Temperature is another critical factor in maintaining water quality. Different fish and plants have different temperature requirements, so it's important to choose species that are compatible with each other and with your local climate. In general, most aquaponic systems do well in temperatures between 70°F and 80°F (21°C to 27°C). If you live in a colder climate, you may need to invest in a heater for your fish tank.

In addition to these basic parameters, you should also keep an eye on dissolved oxygen levels, hardness, and other water quality indicators. Regular testing and adjustments will help you maintain a healthy, balanced ecosystem in your aquaponic system.

Choosing Fish and Plants

Choosing the right fish and plants for your aquaponic system is essential for success. Some fish are better suited to aquaponics than others, and the same goes for plants.

The key is to select species that are compatible with each other and with your local climate.

Fish Species	Size	Growth Rate	Temperature Requirements
Tilapia	Up to 12-14 inches (30-35 cm)	Fast (1-2 inches/month)	75-85°F (24-29°C)
Goldfish	Up to 12 inches (30 cm)	Slow to Moderate	65-75°F (18-24°C)
Koi	Up to 24-36 inches (60-90 cm)	Moderate (2-4 inches/year)	65-75°F (18-24°C)
Catfish	Up to 24 inches (60 cm)	Fast (2-3 inches/month)	70-85°F (21-29°C)

For plants, leafy greens like lettuce, spinach, and kale are ideal for aquaponics because they have relatively low nutrient requirements and grow quickly. Herbs like basil, mint, and parsley also do well in aquaponic systems. If you have more experience, you might try growing fruiting plants like tomatoes, peppers, or strawberries, which have higher nutrient and light requirements.

When selecting plants, consider their growth habits and space requirements. Some plants, like tomatoes, need a lot of vertical space and support, while others, like lettuce, can be grown more densely.

Harvesting and Maintaining Balance

Harvesting in an aquaponic system is a bit different from traditional gardening. Because the system relies on a balance between fish and plants, it's important to harvest carefully to avoid disrupting this balance. Regular, small harvests are usually better than large, infrequent ones.

When harvesting fish, choose individuals that have reached a good size for eating. Be sure to leave enough fish in the tank to maintain nutrient levels for your plants. For plants, harvest the outer leaves of greens and herbs to encourage continued growth. For fruiting plants, pick fruits and vegetables as soon as they are ripe to encourage further production.

Maintaining balance in your aquaponic system involves regular monitoring and adjustments. Keep an eye on water quality, fish health, and plant growth, and make changes as needed. This might involve adjusting feeding rates, changing water more frequently, or adding supplements to your grow bed.

Regular maintenance is also key. Clean filters, check pumps, and inspect your system for any signs of problems. By staying on top of maintenance and monitoring, you can keep your aquaponic system healthy and productive year-round.

Troubleshooting Aquaponics Systems

Even with the best care, problems can arise in an aquaponic system. The key is to identify and address issues quickly to minimize their impact. Common problems include water quality issues, disease, and equipment failures.

If you notice fish acting lethargic or gasping at the surface, it could be a sign of poor water quality. Check ammonia, nitrite, and nitrate levels, and perform a water change if necessary. If plants are yellowing or stunted, it might indicate a nutrient deficiency. Adjust feeding rates or add supplements to your grow bed.

Disease can also be a problem in aquaponic systems. Keep a close eye on your fish and plants for any signs of illness, such as spots, lesions, or discoloration. Quarantine any affected individuals and treat them with appropriate medications or remedies.

Equipment failures can also cause problems. Regularly inspect pumps, filters, and other components for signs of wear or damage. Replace or repair as needed to keep your system running smoothly.

By staying vigilant and proactive, you can quickly address any issues that arise and keep your aquaponic system thriving. With proper care and maintenance, aquaponics can be a rewarding and sustainable way to grow your own food.

CHAPTER 17
Greenhouse Automation

"Technology, when applied thoughtfully, becomes an extension of the gardener's hand, making the care of plants a seamless dance of nature and innovation."

In the ever-evolving world of gardening, greenhouse automation stands as a remarkable testament to how technology can simplify and enhance our growing practices. For many years, I've seen firsthand how the integration of automated systems has transformed traditional gardening into a more efficient and manageable process.

As someone who has juggled multiple responsibilities while maintaining a thriving greenhouse, I can attest to the sheer relief that automation brings. It allows us to focus more on the joy of gardening and less on the minutiae of daily tasks.

Automating Greenhouse Tasks

Automating greenhouse tasks involves using technology to handle routine and repetitive actions, thereby freeing up time and reducing the potential for human error. The concept might sound complex, but it's fundamentally about making our lives as gardeners easier. Whether it's managing irrigation, controlling temperature and humidity, or ensuring adequate ventilation, automation can handle these aspects with precision.

One of the most straightforward applications of automation is in irrigation. Traditionally, watering plants manually can be a time-consuming and inconsistent process. With an automated irrigation system, however, you can schedule watering times and regulate the amount of water delivered, ensuring that each plant receives the right amount without over or under-watering.

For example, drip irrigation systems connected to a timer can be set to water plants at optimal times, like early morning or late evening, which reduces evaporation and maximizes water efficiency.

Another critical task that can be automated is climate control. Maintaining the right temperature and humidity levels is crucial for plant health, but it can be challenging to manage, especially in regions with fluctuating weather conditions.

Automated climate control systems use sensors to monitor the environment inside the greenhouse and make adjustments in real time. For instance, if the temperature rises above a set threshold, an automated system can activate fans or open vents to cool down the space. Conversely, during cooler periods, the system can turn on heaters to maintain a consistent temperature.

Lighting is another area where automation proves invaluable. Plants require different light intensities and durations depending on their growth stage. Automated lighting systems can adjust the brightness and duration of light exposure to suit the needs of your plants. This not only optimizes plant growth but also conserves energy, as lights are only used when necessary.

Types of Automation Systems

There are various types of automation systems available, each designed to address specific aspects of greenhouse management. Understanding these systems will help you decide which ones to incorporate into your setup.

- **Environmental Control Systems**

These systems are the backbone of greenhouse automation, managing temperature, humidity, ventilation, and lighting. They typically include sensors, controllers, and actuators. Sensors measure the environmental parameters, while controllers process this data and send commands to actuators, which then adjust the equipment (like fans, heaters, or lights) accordingly. Advanced systems can even integrate weather forecasts to anticipate and prepare for changes.

- **Irrigation Systems**

As mentioned earlier, automated irrigation systems are designed to deliver water to plants with precision. They can range from simple timers that control sprinklers to sophisticated setups that use moisture sensors and weather data to adjust watering schedules. Some systems also include nutrient dosing capabilities, automatically adding fertilizers to the water based on plant needs.

- **CO2 Enrichment Systems**

Carbon dioxide is essential for photosynthesis, and in controlled environments like greenhouses, CO_2 levels can drop, limiting plant growth. CO_2 enrichment systems automatically monitor and maintain optimal CO_2 levels, ensuring that plants have sufficient resources for healthy growth.

- **Integrated Pest Management (IPM) Systems**

These systems use a combination of monitoring, control, and preventive measures to manage pests and diseases. Automated IPM systems can include pest detection sensors, automated pesticide sprayers, and beneficial insect release systems. They help maintain a balanced ecosystem in the greenhouse, reducing the need for chemical interventions.

Setting Up Automation

Setting up automation in your greenhouse requires careful planning and consideration of your specific needs and budget. Start by assessing the tasks that you find most challenging or time-consuming. For many, this might be climate control or irrigation. Once you've identified your priorities, research the available systems and technologies that can address these needs.

Begin with a basic setup and expand gradually. For example, you might start with automated irrigation and temperature control systems. As you become more comfortable with these technologies, you can add advanced features. It's also crucial to ensure that all components are compatible with each other and can be integrated into a cohesive system.

Installation typically involves placing sensors and controllers in strategic locations around the greenhouse. For irrigation systems, you'll need to install piping and drip lines or sprinklers. Environmental control systems require fans, heaters, vents, and other equipment. Once everything is in place, you'll need to program the controllers with the desired settings and schedules.

Monitoring and Adjusting Systems

Automation doesn't mean you can set everything up and forget about it. Regular monitoring and adjustments are essential to ensure that your systems are functioning correctly and meeting the needs of your plants. Most modern systems come with monitoring software or mobile apps that allow you to check real-time data and system status from anywhere. This is particularly useful for catching issues early, such as a malfunctioning sensor or a blocked irrigation line.

You should also periodically review and adjust your system settings. For example, as the seasons change, you may need to alter temperature and humidity thresholds, irrigation schedules, and light durations. Regular maintenance, such as cleaning filters and checking equipment, is also crucial for preventing system failures and ensuring longevity.

Benefits of Automation

The benefits of greenhouse automation are numerous. First and foremost, it saves time. By automating repetitive tasks, you can focus on more enjoyable aspects of gardening, like choosing new plants or experimenting with different growing techniques. Automation also reduces the risk of human error, ensuring that your plants receive consistent care.

Precise control over watering and nutrient delivery can prevent issues like root rot or nutrient deficiencies. Similarly, automated climate control can protect plants from temperature extremes that might otherwise stress or damage them.

Energy and resource efficiency is another significant benefit. Automated systems can optimize the use of water, electricity, and other resources, reducing waste and lowering costs. For example, automated lighting systems can adjust light levels based on natural sunlight, reducing energy consumption.

Future Trends in Greenhouse Technology

The field of greenhouse automation is constantly evolving, with new technologies and innovations emerging all the time. One exciting trend is the integration of artificial intelligence (AI) and machine learning. These technologies can analyze vast amounts of data to make predictive adjustments and optimize growing conditions.

For example, AI can predict pest outbreaks or nutrient deficiencies based on historical data and sensor readings.

Another trend is the development of more affordable and accessible automation systems. As technology advances, costs are coming down, making it easier for hobbyist gardeners to adopt automation. Wireless and modular systems are also becoming more popular, allowing for easier installation and expansion.

Finally, there's a growing focus on sustainability and eco-friendliness in greenhouse automation. This includes using renewable energy sources, like solar panels, to power systems, as well as developing more energy-efficient equipment and practices.

Greenhouse automation is a powerful tool that can transform your gardening experience. Whether you're a seasoned gardener or a beginner, these technologies can help you create a thriving, efficient, and enjoyable greenhouse. As we continue to explore and embrace these innovations, the future of gardening looks brighter and more sustainable than ever.

PART 6: GREENHOUSE GARDENING BEYOND THE BASICS

CHAPTER 18

Extending Your Growing Season

"The best way to predict the future is to create it—by extending your growing season, you're not just making plans, you're making possibilities."

One of the greatest joys of greenhouse gardening is the ability to extend the growing season beyond the traditional limits of nature. Over the years, I've found that the real magic happens when you harness the power of your greenhouse to maximize harvests all year round.

Whether you're new to greenhouse gardening or a seasoned veteran, the techniques we'll cover in this chapter will help you push the boundaries of what's possible.

Techniques for Maximizing Harvests

Extending your growing season in a greenhouse isn't just about keeping your plants alive longer—it's about maximizing what you get from every inch of space and every minute of light. When I first started experimenting with extending my growing season, I realized that success hinges on a blend of strategic planning and a willingness to get creative.

One of the most effective ways to maximize your harvests is through succession planting. By carefully planning and timing your plantings, you can ensure that as one crop finishes, another is ready to take its place. This method requires a bit of foresight and organization, but the payoff is a continuous supply of fresh produce.

Another technique that has worked wonders for me is intercropping. This involves planting different crops together in a way that they complement each other and make efficient use of space and resources. For example, you might plant fast-growing lettuce alongside slower-growing tomatoes.

The lettuce can be harvested before the tomatoes get too big, making the most of the space. This not only maximizes your yield but also helps with pest control and soil fertility, as different plants have different nutrient needs and pest attractants.

Row covers and cloches are also invaluable tools in extending your growing season. These simple yet effective tools can help protect your plants from the cold, wind, and pests. Row covers, for instance, can be used to trap heat and moisture around plants, effectively creating a mini-greenhouse effect.

Cloches, on the other hand, are individual coverings for plants that provide a similar benefit. I've found that using these can mean the difference between a bountiful harvest and a disappointing one, especially in unpredictable spring and fall weather.

But it's not just about keeping plants warm. Maximizing your harvests also involves optimizing the growing conditions inside your greenhouse. This means paying close attention to the microclimates you create. For example, placing heat-loving plants in the sunniest, warmest spots and cooler-season crops in shadier, cooler areas can make a significant difference.

In my experience, keeping a garden journal is an invaluable tool in this process. By recording what you've planted, where, and when, as well as the conditions and outcomes, you can fine-tune your approach year after year. It's a bit like being a detective, piecing together clues to find out what works best in your unique greenhouse environment.

Lastly, don't underestimate the power of good soil management. Healthy, nutrient-rich soil is the foundation of any successful garden. In a greenhouse, where space is limited and plants are often in close quarters, maintaining soil health is even more critical. Regularly adding compost and other organic matter, rotating crops, and using cover crops can help keep your soil in top condition.

Using Season Extenders

Season extenders are a greenhouse gardener's secret weapon. They allow you to push the boundaries of the traditional growing season, letting you start planting earlier in the spring and continue harvesting well into the fall or even winter. When I first started using season extenders, it felt like unlocking a whole new world of possibilities. Suddenly, I could grow crops that I had only dreamed of cultivating in my climate.

One of the simplest and most effective season extenders is the humble row cover. These lightweight fabrics can be draped over plants to provide a layer of protection from the cold. They work by trapping heat and humidity, creating a microenvironment that can be several degrees warmer than the surrounding air.

This small increase in temperature can be all your plants need to survive a chilly night or get a head start in the early spring. I remember the first time I used row covers to protect a bed of tender spinach and lettuce. Not only did they survive a late frost, but they thrived, giving me a much earlier harvest than I expected.

Cold frames are another fantastic tool. Essentially mini-greenhouses, cold frames are low, box-like structures with a transparent top that can be opened and closed. They're perfect for hardening off seedlings, starting early crops, or extending the harvest of cold-tolerant vegetables like kale and spinach.

Building my first cold frame was a game-changer. It was a simple wooden frame with an old window on top, but it allowed me to grow delicious greens through the winter, something I had never thought possible before.

For those looking to take it a step further, hotbeds are an exciting option. These are essentially cold frames with a heat source, traditionally created with a layer of decomposing manure beneath the soil. As the manure breaks down, it generates heat, warming the soil and air inside the hotbed.

This technique can give you a significant temperature boost, perfect for starting early seedlings or growing more heat-loving plants in cooler seasons. I've experimented with hotbeds over the years, and the results have been incredible. The extra warmth can turn your greenhouse into a tropical oasis, even in the dead of winter.

Then, there are more high-tech solutions like heated mats and grow lights. While they require an initial investment and a bit more management, they can provide consistent, controllable heat and light, allowing you to grow just about anything, anytime.

I've used heated mats to start seedlings early in the season, ensuring a steady supply of transplants ready to go into the ground as soon as the weather permits. Grow lights, meanwhile, have been a lifesaver for extending the day length during the dark winter months, keeping my plants growing strong.

Using season extenders effectively often involves a bit of trial and error. You have to experiment with different methods and combinations to see what works best for your particular setup and climate. But that's part of the fun of greenhouse gardening—always learning and adapting, pushing the limits of what you can grow.

Selecting Short-Growing Varieties

When it comes to extending your growing season, one of the most effective strategies is choosing the right plant varieties. Not all plants are created equal, especially when it comes to how long they take to mature. Selecting short-growing varieties can significantly impact your greenhouse's productivity, allowing you to fit more growing cycles into a single season. This strategy is particularly useful if you're working with a limited timeframe due to climate constraints.

I learned the value of short-growing varieties early in my greenhouse gardening journey. It's a simple concept but incredibly powerful: plants that mature faster can be harvested sooner, freeing up space for another crop. For instance, radishes are a perfect example of a quick-growing crop. Some varieties can be ready to harvest in as little as 25 days. They're a fantastic option for early spring or late fall, filling in gaps when other crops are still getting started or winding down.

Lettuce and other leafy greens also come in many short-growing varieties. I've found that baby leaf varieties are especially useful in a greenhouse setting. These greens can be harvested in just a few weeks, providing a quick turnaround from planting to plate.

They're also versatile, fitting well into both direct-seed and transplant schedules. In my greenhouse, I often use them as a filler crop, planting them between longer-season vegetables like tomatoes or peppers. This way, I'm always making the most of my space.

Peas and beans, particularly the bush types, are another great option. They tend to mature faster than their pole counterparts and can be grown in succession throughout the season. I've had great success with early-maturing pea varieties, enjoying fresh pods while the cool weather lasts. Then, as the temperature rises, I switch to bush beans, which thrive in the warmer conditions of mid-summer.

Choosing short-growing varieties also allows you to experiment with multiple planting cycles within the same season. This is especially useful in a greenhouse, where you can control the environment to some extent.

For example, you can start with a quick crop of spinach in early spring, followed by a heat-tolerant crop like basil in the summer, and finish with another round of spinach or kale as the temperatures cool again. This cycle not only maximizes your harvests but also keeps your gardening experience fresh and exciting, with new crops constantly coming in.

Another aspect to consider is the use of determinate vs. indeterminate varieties, especially with crops like tomatoes and peppers. Determinate varieties, which grow to a certain size and then stop, tend to have a shorter growing period and can produce a concentrated harvest.

This can be ideal for greenhouse gardeners looking to maximize space and time. Indeterminate varieties, on the other hand, grow and produce fruit over a longer period. While they might not fit into a tight growing schedule as neatly, they offer the benefit of extended harvests, which can be managed with careful planning.

When selecting short-growing varieties, it's important to read the seed packet information carefully. Look for days to maturity, and consider how these fit into your overall gardening plan. It's also worth experimenting with different varieties to find what works best in your specific greenhouse conditions. Over the years, I've discovered some surprising winners that thrive in my particular setup, often defying my initial expectations.

Managing Light and Temperature

Managing light and temperature is one of the most critical aspects of extending your growing season in a greenhouse. These two factors, more than anything else, determine the health and productivity of your plants. Over the years, I've learned that getting the balance right can make the difference between a bountiful harvest and a disappointing yield. Let's dive into some practical strategies to optimize these essential elements.

First and foremost, understanding the natural light cycle and how it changes with the seasons is crucial. At peak times of summer, your greenhouse might receive ample sunlight, sometimes even too much, which can lead to overheating. In contrast, during the winter months, daylight hours shorten, and the intensity of the sunlight diminishes. To manage these fluctuations, consider using shading materials or shade cloth during the hottest months.

These can help diffuse intense sunlight, preventing leaf scorch and overheating. I remember a summer when I neglected to put up my shade cloth early enough, and the result was a batch of tomato plants that struggled in the scorching heat. Since then, I've been diligent about putting up the shade cloth as soon as the temperatures start to climb.

LED grow lights offer energy efficiency and flexibility in providing the specific light spectrum that plants need. When choosing LED grow lights, consider the following spectrum features:

Full Spectrum

- Description: Mimics natural sunlight and supports all stages of plant growth.
- Benefits: Provides a balanced spectrum of light, including blue, red, and far-red wavelengths, making it suitable for both vegetative growth and flowering.
- Usage: Ideal for general-purpose use, including growing a wide range of plants from seedlings to mature crops.

Blue Light (450-470 nm)

- Description: Focuses on the blue part of the spectrum.
- Benefits: Promotes vegetative growth, enhances leaf development, and helps regulate plant growth patterns.

- Usage: Best used during the vegetative stage of plant growth to support strong, healthy foliage.

Red Light (620-660 nm)

- Description: Emphasizes the red part of the spectrum.
- Benefits: Encourages flowering, fruiting, and overall plant maturity. Essential for the blooming phase.
- Usage: Important for the flowering stage, promoting the production of flowers and fruit.

Far-Red Light (700-740 nm)

- Description: Falls beyond the red spectrum.
- Benefits: Influences flowering and can extend the photoperiod, which can be useful for plants that require longer light periods.
- Usage: Can be used in conjunction with red light to optimize flowering and fruiting.

Cool White Light

- Description: Includes a broad spectrum of light with a cooler temperature.
- Benefits: Provides additional light coverage and supports all stages of growth.

- Usage: Useful for providing a broad spectrum of light, often included in full-spectrum LED lights.

Warm White Light

- Description: Provides a warmer light with a focus on the red and orange spectrum.
- Benefits: Supports flowering and fruiting, similar to natural sunlight in the late afternoon.
- Usage: Often combined with other spectrums to mimic natural light conditions and promote flowering.

Key Considerations

- PAR (Photosynthetically Active Radiation): Look for LED grow lights with a high PAR rating to ensure they provide sufficient light intensity for photosynthesis.
- Adjustability: Some LED grow lights offer adjustable spectrums, allowing you to tailor the light to different growth stages.
- Coverage Area: Choose lights that provide adequate coverage for the size of your growing area to ensure even light distribution.

I've used grow lights to great effect in my greenhouse, particularly for starting seeds early in the spring. The extra light helps seedlings develop strong stems and leaves, readying them for transplant when the outdoor conditions are right.

As discussed in previous chapters, temperature control goes hand-in-hand with managing light. I've found that using a combination of passive solar heating—such as thermal mass (like water barrels) to store and release heat—and a backup heater works well.

The thermal mass absorbs heat during the day and releases it slowly at night, helping to moderate temperature swings. On particularly cold nights, the backup heater kicks in to maintain a consistent temperature, ensuring my more delicate plants stay safe.

Don't forget about the importance of insulating your greenhouse, particularly in winter. Insulation can help retain the heat generated during the day and reduce the amount of additional heating needed at night. This can be achieved by using bubble wrap or other insulating materials on the walls and roof. I've wrapped my greenhouse in bubble wrap for several winters, and the difference it makes in maintaining a stable temperature inside is remarkable.

Another useful technique is to divide your greenhouse into different zones, each with its own microclimate. This can be done with curtains or partitions, allowing you to create warmer and cooler areas within the same space.

For instance, you might keep more heat-sensitive plants near the ventilated areas while grouping cold-hardy crops in the cooler zones. This zoning approach has allowed me to grow a diverse range of plants simultaneously, each in the conditions that suit them best.

Lastly, always monitor your greenhouse environment. Using thermometers, hygrometers, and light meters can help you keep track of the conditions and make adjustments as needed.

Many gardeners, including myself, find that maintaining a greenhouse journal helps to track these variables over time, allowing for more informed decisions in future seasons.

Timing Your Plantings

Timing your plantings in a greenhouse setting is both an art and a science. It involves careful planning, observation, and a bit of intuition. The goal is to synchronize your planting schedule with the changing environmental conditions inside and outside the greenhouse. This allows you to make the most of the growing season and ensure a continuous harvest.

One of the first things to consider is the natural growing season of each plant. Understanding whether a plant is in the cool season or warm season will help you determine the best times to start and finish each crop. Cool-season crops like lettuce, spinach, and peas thrive in the cooler temperatures of early spring and fall.

Warm-season crops like tomatoes, peppers, and cucumbers prefer the heat of summer. In a greenhouse, you can start cool-season crops earlier and keep them going later, effectively doubling their usual growing season. I've enjoyed fresh spinach well into December, long after outdoor gardens have succumbed to frost, simply by planting at the right times.

Succession planting is another key strategy to maximize harvests, as I mentioned before in “Techniques for Maximizing Harvest”. This strategy involves planting a new crop as soon as another one finishes.

The idea is to have seedlings ready to go as soon as space opens up. Over the years, I've developed a system where I always have seedlings in trays, ready to replace harvested crops. This not only maximizes space but also ensures a steady supply of fresh produce.

Staggered planting is also a valuable technique. By planting the same crop at two or three-week intervals, you can extend the harvest period. This is particularly useful for crops like lettuce, which can bolt quickly in warm temperatures.

Staggered planting ensures that you always have fresh, tender greens available. I've done this with great success, allowing me to enjoy fresh salads throughout the growing season without a break.

When it comes to timing, it's also crucial to consider the specific conditions within your greenhouse. Each greenhouse has its own unique microclimate, influenced by factors like orientation, insulation, and the presence of heat sinks.

Observing these conditions and how they change over the course of the day and season will inform your planting schedule. In my own greenhouse, I've noticed that certain corners stay cooler or warmer depending on the time of year. I've used this knowledge to my advantage, placing crops in the spots that suit them best.

Day length is another critical factor. Some plants require a certain amount of daylight to flower and produce fruit. This is where supplemental lighting can come in handy, especially during the short days of winter.

For example, I've successfully grown cherry tomatoes in the dead of winter by providing additional light. The key is to simulate the long days of summer, ensuring the plants get enough light to thrive. This kind of planning allows you to grow summer crops out of season, providing fresh produce when it's typically unavailable.

Frost dates are also crucial when planning your plantings. Knowing the average last frost date in spring and the first frost date in fall helps you avoid planting too early or too late. In a greenhouse, you have more flexibility, but it's still important to pay attention to these dates.

For tender plants, like tomatoes or peppers, I always make sure to have backup heating options ready if an unexpected frost threatens. Conversely, in the fall, I use frost blankets and row covers to extend the growing season of my cold-sensitive plants.

Lastly, consider the lifecycle of each plant. Some crops, like lettuce or radishes, mature quickly and can be planted and harvested multiple times in a season. Others, like tomatoes or peppers, have a longer growing cycle and may only yield one or two harvests.

Understanding these lifecycles helps you plan your planting and harvesting schedule more effectively. Over time, you'll develop a rhythm that works best for your specific greenhouse setup and your goals.

Protecting Against Frost

Frost is one of the biggest challenges for greenhouse gardeners, especially if you live in a region with a long winter or unpredictable spring weather. Even a light frost can damage or kill tender plants, so it's crucial to have strategies in place to protect your crops. When I first started gardening in a greenhouse, I quickly learned that frost protection is not just a matter of keeping the plants warm—it's about anticipating the cold and acting proactively.

One of the simplest and most effective methods for frost protection is using row covers. These lightweight fabrics can be draped over plants to provide an extra layer of insulation. They trap the heat that radiates from the soil, keeping the air around the plants a few degrees warmer.

I've found that row covers are incredibly versatile. They can be used directly on plants or supported by hoops to create mini-tunnels. In early spring, I often set up these tunnels over my cold-hardy crops, giving them a head start and protecting them from late frosts.

Frost blankets are another essential tool. They are similar to row covers but thicker and more insulating. These are particularly useful for protecting tender plants during unexpected cold snaps. I remember one particularly cold spring when an unexpected frost threatened my newly planted tomatoes. I quickly covered them with frost blankets, securing the edges with rocks to keep out the chill. The extra effort paid off; the plants survived and thrived, producing a bountiful harvest later in the season.

Another technique I've used is mulching. Applying a thick layer of mulch around the base of plants helps retain soil moisture and warmth. Mulch acts as an insulating layer, protecting the roots from freezing. I've used straw, leaves, and even wood chips, depending on what I have available.

In addition to frost protection, mulching also suppresses weeds and improves soil structure, making it a valuable practice year-round.

For more high-tech solutions, consider using heaters. I've set up a few small heaters in my greenhouse, which are connected to a thermostat that turns them on and off as needed. This setup ensures that the temperature stays within a safe range without constant supervision.

Water barrels are a popular choice; they absorb the sun's heat during the day and radiate it back out as temperatures drop. I've lined the north wall of my greenhouse with water barrels painted black. This not only helps with temperature regulation but also provides a handy source of water for the plants.

Another important aspect of frost protection is timing your plantings carefully. In the fall, I often monitor the weather forecast closely, ready to cover or harvest crops if an early frost is predicted. It's a balancing act between extending the growing season and protecting the plants.

Lastly, consider the layout of your greenhouse. Frost often settles in low-lying areas, so it's a good idea to keep sensitive plants on higher benches or shelves. I've arranged my greenhouse so that the most delicate plants are elevated, reducing the risk of frost damage. This simple adjustment can make a significant difference, especially on cold nights.

Protecting against frost requires a combination of strategies and a proactive approach. By using row covers, frost blankets, mulching, heaters, and thermal mass, you can create a more stable environment for your plants. And by paying close attention to weather patterns and frost dates, you can take timely actions to safeguard your crops. With these measures in place, you can confidently extend your growing season and enjoy fresh produce well beyond the traditional limits.

CHAPTER 19

Sustainable Practices for Your Greenhouse

"The true measure of our progress is not how much we grow, but how sustainably we grow."

I've learned that sustainability isn't just a trend; it's a commitment to nurturing the environment while reaping the benefits of a thriving garden. Over the years, I've seen firsthand how adopting eco-friendly practices can transform a greenhouse into a more productive and harmonious space. Here, we'll explore the journey of making your greenhouse greener.

From reducing environmental impact to embracing organic pest control and energy-efficient practices, I'll share the methods that have worked for me and countless other gardeners. Together, we'll delve into practical strategies that not only help protect our planet but also make our gardening endeavors more rewarding and fulfilling.

Organic Pest Control

Organic pest control is a cornerstone of sustainable greenhouse management. By focusing on natural methods to manage pests, you not only protect your plants but also contribute to a healthier environment.

- **Understanding Beneficial Insects**

One of the most effective ways to control pests organically is by introducing beneficial insects into your greenhouse. Ladybugs, lacewings, and predatory mites are excellent examples of beneficial insects that prey on common greenhouse pests such as aphids, spider mites, and whiteflies. By creating a habitat for these beneficial insects, you encourage their presence and enhance their pest-fighting abilities.

- **Using Natural Pesticides**

Natural pesticides can be an effective alternative to synthetic chemicals. Neem oil, for example, is a versatile natural pesticide that disrupts the life cycle of pests and acts as a repellent. Similarly, insecticidal soap, made from natural plant oils and fats, can control soft-bodied insects like aphids and whiteflies without harming beneficial insects.

- **Implementing Physical Barriers**

Physical barriers are another effective method of organic pest control. Row covers and netting can protect plants from flying pests while allowing light and water to reach the plants. These barriers are especially useful for shielding young seedlings from insect infestations and preventing pests from laying eggs on your plants.

- **Encouraging Natural Predators**

Encouraging natural predators is a key component of organic pest control. Birds, frogs, and other wildlife can play a significant role in controlling pest populations. To attract these natural predators, create a habitat that provides food, water, and shelter. Install bird feeders, provide water sources, and plant a variety of flowering plants that attract insects and other wildlife.

- **Maintaining Plant Health**

Maintaining overall plant health is crucial for preventing pest problems. Healthy plants are more resilient to pest attacks and diseases. Ensure that your plants receive adequate nutrients, water, and light. Regularly inspect your plants for signs of stress or disease, and address any issues promptly to prevent pests from taking hold.

- **Implementing Companion Planting**

Companion planting involves growing certain plants together to repel pests and enhance plant health. For example, planting marigolds alongside tomatoes can help deter nematodes and other soil-borne pests. Similarly, garlic and chives can repel aphids and other insects when planted near susceptible plants. Research and plan your plant combinations to maximize the benefits of companion planting in your greenhouse.

- **Monitoring and Adjusting**

Regular monitoring is essential for effective organic pest control. Keep a close eye on your plants for signs of pest activity and take action as soon as you notice any issues. Adjust your pest control strategies as needed based on the type and severity of the infestation.

Water Conservation

Water conservation is a crucial aspect of sustainable greenhouse management. Efficient water use not only helps to preserve this vital resource but also contributes to the overall health and productivity of your greenhouse.

- **Implementing Efficient Irrigation Systems**

One of the most effective ways to conserve water in your greenhouse is by using efficient irrigation systems. Drip irrigation, soaker hoses or subsurface irrigation are good choices to reduce water wastage.

- **Utilizing Rainwater Harvesting**

Rainwater harvesting is a valuable technique for conserving water in your greenhouse. By collecting and storing rainwater, you can reduce your reliance on municipal water sources. Set up rain barrels or storage tanks to capture rainwater from your greenhouse roof. This water can then be used for irrigation, helping you save on water costs and reduce your environmental footprint.

- **Reducing Evaporation**

Reducing evaporation is another key aspect of water conservation. Mulching is an effective method for minimizing evaporation from the soil. Apply organic mulch, such as straw or wood chips, around your plants to retain moisture and regulate soil temperature.

- **Monitoring Soil Moisture**

Monitoring soil moisture levels is essential for efficient water use. Overwatering not only wastes water but can also lead to root rot and other plant issues. Use moisture sensors or a simple soil moisture meter to gauge the moisture levels in your greenhouse. Adjust your irrigation schedules based on the readings to ensure that you're providing just the right amount of water.

- **Recycling Water**

Recycling water is another effective way to conserve resources. Implement a system to capture and reuse runoff water from irrigation or rainwater collection. This water can be filtered and used for irrigation, reducing the amount of freshwater needed. Additionally, consider using greywater systems to recycle water from sinks or showers, provided it is properly treated and safe for plant use.

- **Educating and Involving Others**

Educating others about the importance of water conservation can further enhance your efforts. Share your knowledge with fellow gardeners, greenhouse staff, or community members. Encourage them to adopt water-saving practices and implement efficient irrigation systems in their own gardens. By working together, we can collectively reduce water consumption and promote sustainable practices.

- **Regular Maintenance**

Regular maintenance of your irrigation systems is crucial for ensuring optimal water conservation. Inspect your irrigation equipment regularly for leaks, clogs, or damage. Additionally, schedule periodic checks to ensure that your irrigation system is functioning efficiently and effectively.

Sustainable Building Materials

When it comes to constructing or renovating your greenhouse, choosing sustainable building materials is a key aspect of reducing your environmental impact. By selecting materials that are eco-friendly and responsibly sourced, you can create a greenhouse that is both efficient and environmentally conscious. In this section, I'll explore various sustainable building materials and practices that will help you build a greener greenhouse.

- **Selecting Recycled Materials**

Recycled materials are an excellent choice for sustainable construction. Recycled steel, for example, is a durable and environmentally friendly option for greenhouse frames. Using recycled steel reduces the need for new raw materials and helps to minimize waste. Similarly, consider using reclaimed wood for construction projects, such as benches or shelving, to further reduce your environmental footprint.

Repurpose leftover materials from previous projects or source second-hand items for construction. Additionally, a waste management plan should be implemented to ensure that construction waste is sorted and recycled appropriately.

- **Using Sustainable Sourcing**

When selecting building materials, it's important to choose products that are sustainably sourced. Look for materials that come from renewable resources or are certified by organizations such as the Forest Stewardship Council (FSC). For instance, bamboo is a rapidly renewable resource that can be used for various construction purposes, including flooring and supports.

- **Incorporating Energy-Efficient Glazing**

Energy-efficient glazing is a crucial component of sustainable greenhouse construction. Opt for glazing materials that offer excellent insulation properties, such as polycarbonate panels or double-glazed glass. These materials help to maintain a stable internal temperature, reducing the need for heating and cooling and improving overall energy efficiency.

- **Exploring Green Roof Options**

Green roofs are an innovative and sustainable option for greenhouse construction. A green roof involves planting vegetation on the roof of your greenhouse, which provides additional insulation, reduces heat absorption, and improves air quality. This practice not only benefits your greenhouse environment but also contributes to urban green spaces and biodiversity.

- **Utilizing Low-Impact Finishes**

Low-impact finishes, such as natural paints and non-toxic sealants, are essential for maintaining a sustainable greenhouse environment. Choose finishes that are free from harmful chemicals and have minimal impact on indoor air quality. Additionally, consider using low-VOC (volatile organic compounds) paints and finishes to ensure a healthier environment for both you and your plants.

- **Designing for Longevity**

Designing your greenhouse with longevity in mind is a key aspect of sustainability. Opt for materials and construction methods that are durable and require minimal maintenance. Investing in high-quality, long-lasting materials will reduce the need for frequent repairs or replacements, ultimately contributing to a more sustainable and resource-efficient greenhouse.

Energy-Efficient Practices

Energy-efficient practices are essential for maintaining a sustainable greenhouse environment. By adopting strategies that minimize energy consumption, you can reduce operational costs and lower your greenhouse's carbon footprint. In this section, I'll revise various energy-efficient practices and technologies that will help you create a more sustainable and cost-effective greenhouse.

- **Optimizing Heating and Cooling Systems**

Efficient heating and cooling systems are crucial for maintaining optimal growing conditions while minimizing energy use. For heating, consider using energy-efficient options such as infrared heaters. Geothermal heating systems can harness the earth's natural heat to maintain a stable temperature with minimal energy input.

For cooling, implement natural ventilation techniques, such as roof vents and side vents. Shade cloths and thermal screens can also help regulate temperature reducing heat buildup inside the greenhouse.

- **Utilizing Passive Solar Design**

Passive solar design involves harnessing natural sunlight to heat and light your greenhouse. Position your greenhouse to maximize exposure to the sun, and use materials with good thermal mass, such as concrete or brick, to store and release heat. Install large, south-facing windows or skylights to allow sunlight to enter and warm the greenhouse naturally.

- **Implementing Energy-Efficient Lighting**

LED grow lights are an excellent choice for providing the necessary light spectrum for plant growth while consuming significantly less energy than traditional lighting options. Additionally, use timers and dimmers to regulate light intensity and duration, ensuring that your plants receive optimal lighting while minimizing energy use.

- **Monitoring and Controlling Energy Use**

Regularly monitoring and controlling energy use is essential for identifying areas for improvement and optimizing efficiency. Install energy meters or sensors to track your greenhouse's energy consumption and identify any inefficiencies. Use this data to make informed decisions about adjustments or upgrades to your energy systems.

- **Incorporating Smart Technology**

Smart technology can enhance energy efficiency by automating various greenhouse functions. For example, smart thermostats can regulate heating and cooling based on real-time data, while automated shading systems can adjust according to sunlight intensity. By integrating smart technology into your greenhouse, you can optimize energy use and reduce waste.

- **Conducting Energy Audits**

Conducting regular energy audits can help you assess the efficiency of your greenhouse's systems and identify opportunities for improvement. An energy audit involves evaluating your greenhouse's energy consumption, identifying areas of inefficiency, and recommending solutions to enhance performance. Implementing the recommendations from your energy audit can lead to significant energy savings and improved sustainability.

- **Educating and Training**

Educating and training your greenhouse staff or family members on energy-efficient practices is essential for maintaining a sustainable environment. Share knowledge about energy conservation techniques, efficient equipment uses, and the benefits of adopting sustainable practices.

Recycling and Waste Management

Recycling and waste management are crucial aspects of maintaining a sustainable greenhouse. By implementing effective strategies for handling waste, you can reduce environmental impact, conserve resources, and create a more efficient and eco-friendly greenhouse environment.

- **Implementing a Composting System**

Composting is a highly effective way to recycle organic waste and create nutrient-rich soil for your greenhouse. Set up a composting system to process plant debris, kitchen scraps, and other organic materials. Composting not only reduces waste but also enriches the soil with essential nutrients, promoting healthy plant growth.

- **Sorting and Recycling Materials**

Sorting and recycling materials is an essential step in waste management. Set up designated bins for different types of waste, such as plastics, glass, paper, and metal. Educate your staff or family members on proper sorting techniques and ensure that recyclables are clean and free from contaminants. By recycling materials, you can reduce the amount of waste sent to landfills and support the circular economy.

- **Reducing Single-Use Plastics**

Reducing single-use plastics is an important aspect of sustainable waste management. Avoid using disposable plastic items, such as pots and trays, and opt for reusable alternatives instead. Using biodegradable or compostable materials whenever possible helps to minimize pollution and supports a healthier environment.

- **Regular Waste Audits**

Conducting regular waste audits can help you assess the effectiveness of your recycling and waste management practices. A waste audit involves analyzing the types and quantities of waste generated, identifying areas for improvement, and implementing strategies to reduce waste. Use the findings from your waste audit to make informed decisions and enhance your waste management practices.

- **Implementing a Waste Reduction Plan**

Developing and implementing a waste reduction plan is essential for achieving long-term sustainability goals. Outline specific strategies and goals for reducing waste, recycling materials, and managing compost. Monitor progress regularly and make adjustments as needed to ensure that your waste reduction efforts are effective and aligned with your sustainability objectives.

CHAPTER 20
Expanding Your Greenhouse Operations

"Every great gardener starts with a small plot, but it's the ambition to grow that truly cultivates success."

When I first set up my greenhouse, I was content with a modest variety of vegetables and herbs, simply relishing the joy of gardening. But as the seasons passed, I found myself yearning to explore new horizons and expand my work.

Expanding your greenhouse operations can be an exciting and rewarding endeavor, whether you're a hobbyist looking to diversify your crops or an entrepreneur aiming to turn your passion into a profitable business.

Diversifying Your Crops

Expanding your crop variety is one of the most rewarding ways to utilize your greenhouse space. I remember when I first ventured beyond my usual tomatoes and cucumbers. I decided to try my hand at growing exotic herbs and edible flowers. The transition wasn't without its challenges, but it opened up a whole new world of possibilities and flavors.

Diversifying your crops involves more than just picking new seeds; it requires understanding the specific needs of each plant type. Different crops have varying requirements for light, temperature, humidity, and soil nutrients. For instance, while tomatoes thrive in warmer conditions, leafy greens like spinach prefer cooler environments. In my experience, setting up different zones within the greenhouse helped manage these diverse needs effectively.

To start diversifying, you might want to experiment with high-value crops like strawberries, peppers, or even rare herbs. These not only add variety but can also command higher prices if you decide to sell your produce. I found that growing herbs like basil, mint, and rosemary not only provided fresh flavors for my kitchen but also attracted local chefs interested in buying fresh, aromatic herbs directly from the source.

Another aspect of diversifying is staggered planting. By planting crops with different growth cycles, you can ensure a continuous harvest. This technique has been particularly useful for me in maintaining a steady supply of fresh produce throughout the year. For example, while my winter spinach is reaching maturity, I've already started seedlings for spring crops like peas and lettuce.

When diversifying, consider the market demand if you're planning to sell your produce. Research local farmers' markets, grocery stores, and restaurants to understand what products are in demand.

I remember attending local farmers' markets and chatting with customers and vendors to get a feel for what was popular. This not only helped me choose crops that would sell well but also gave me valuable insights into pricing and marketing strategies.

Lastly, keep a detailed record of your planting schedules, harvest dates, and crop performance. This data will be invaluable as you refine your operations. I've kept a gardening journal for years, noting down everything from seed varieties and planting dates to pest issues and yields. This practice has helped me make informed decisions and improve my crop diversity strategy over time.

Using the Greenhouse for Non-Food Plants

While growing food crops is the primary use of many greenhouses, don't overlook the potential of non-food plants. Ornamental plants, medicinal herbs, and even tropical species can thrive in a controlled greenhouse environment. My journey into non-food plants began with a simple fascination with orchids. These beautiful yet delicate plants require specific conditions that my greenhouse could easily provide.

Ornamentals like orchids, ferns, and succulents can be a delightful addition to your greenhouse, offering aesthetic appeal and potential commercial opportunities. I started with a few orchid varieties, experimenting with different humidity levels and lighting conditions. Over time, I expanded to include ferns and air plants, which added a lush, green ambiance to the greenhouse.

Medicinal herbs are another excellent choice for non-food plants. Herbs like echinacea, chamomile, and lavender not only have therapeutic properties but also add diversity to your greenhouse. I found growing medicinal herbs particularly rewarding, as they require minimal space and can be processed into various products like teas, oils, and salves.

Sharing these products with friends and family has been a joy, and they've often encouraged me to consider selling them locally.

Tropical plants, while more demanding in terms of temperature and humidity, can also be a fascinating addition. I've dabbled in growing tropical fruits like pineapples and passionfruit, which, while challenging, have been incredibly rewarding. The sight of a pineapple growing in my greenhouse in a temperate climate never fails to amaze visitors.

If you're interested in non-food plants for decorative purposes, consider seasonal plants like poinsettias for the winter holidays or lilies for the spring. These can be sold during specific seasons, providing an additional revenue stream. I remember one holiday season when I decided to grow poinsettias. The vibrant red and green plants were a hit at local markets, and I even received orders from local businesses looking to decorate their offices.

Growing non-food plants also allows for creative arrangements and displays. I've enjoyed creating themed sections in my greenhouse, like a tropical corner with palm trees and hibiscus or a serene space filled with fragrant lavender and rosemary. These themed areas not only look beautiful but also serve as a great way to showcase the diversity of plants you can grow.

Greenhouse as a Business

Turning your greenhouse hobby into a business can be both exciting and daunting. When I first considered this step, the idea of selling my produce and plants seemed like a natural progression. However, there are several key factors to consider to ensure your venture is successful.

First, understand your market. Whether you're selling at farmers' markets, supplying restaurants, or offering ornamental plants, it's crucial to know who your customers are and what they're looking for. My early experiences selling at local markets taught me the importance of customer feedback. I learned to adapt my offerings based on demand, which helped build a loyal customer base.

Next, think about branding and marketing. Your greenhouse products need a unique selling proposition (USP). For me, it was the freshness and variety of produce, along with the option to buy unique herbs and plants not readily available in stores. Creating a brand story and logo helped establish my presence in the market. I also used social media platforms to showcase my products, share gardening tips, and connect with potential customers.

Pricing is another critical aspect. It's important to set prices that reflect the quality and uniqueness of your products while remaining competitive. Initially, I struggled with pricing my products fairly. I didn't want to undercut the market, but I also needed to cover my costs and make a profit. After some research and adjustments, I found a pricing strategy that worked. I priced premium products like exotic herbs slightly higher due to their rarity and the effort involved in growing them.

Logistics and distribution can be challenging, especially if you're dealing with perishable items. I decided to start small, focusing on local markets and direct sales to restaurants. This approach minimized the logistical challenges and allowed me to maintain a high standard of quality. Over time, as I expanded, I explored delivery options and partnerships with local stores.

Legal considerations are also a vital part of setting up a greenhouse business. You'll need to research local regulations regarding agricultural production, sales, and possibly organic certification if you're marketing your produce as organic. I consulted with a local business advisor to ensure I was compliant with all necessary regulations, including food safety standards and tax requirements.

Lastly, don't underestimate the value of excellent customer service. Building relationships with your customers can lead to repeat business and positive word-of-mouth. I always made it a point to engage with customers, answer their questions, and even give them a tour of my greenhouse when possible. This personal touch helped establish trust and loyalty.

Marketing Greenhouse Products

Marketing is essential to the success of your greenhouse business. It's about more than just advertising; it's about telling your story and connecting with your audience. I found that sharing my journey as a gardener, the challenges I've faced, and the successes I've enjoyed resonated with people.

One of the first steps in marketing is identifying your target audience. Are you selling to health-conscious consumers, chefs, or plant enthusiasts? Knowing your audience helps tailor your marketing efforts. For me, a mix of local residents looking for fresh produce and restaurants interested in unique herbs and vegetables made up the bulk of my customers.

Creating a strong online presence is crucial in today's digital age. I set up a simple website showcasing my products, including an online store for easy ordering. Social media platforms like Instagram and Facebook were invaluable for sharing photos of my greenhouse, plants, and the gardening process. Engaging with followers through posts, stories, and live videos helped create a community around my brand.

I also utilized local media outlets to spread the word. Press releases about new products, seasonal offerings, or special events caught the attention of local newspapers and magazines. This free publicity was a great way to reach a broader audience. One year, I hosted an open house event at my greenhouse, inviting the community to explore and purchase directly from the source. The local newspaper covered the event, which brought in many new customers.

Participating in farmers' markets and local fairs is another effective marketing strategy. These venues provide a platform to showcase your products, engage with potential customers, and receive immediate feedback. I often brought samples of my produce for people to taste, which proved to be a great way to entice them to buy. The direct interaction with customers at these events also helped build brand recognition and loyalty.

Offering workshops or gardening classes can also be an excellent way to market your greenhouse. I started offering workshops on herb gardening, which not only generated additional income but also positioned me as an expert in the community. These workshops often led to increased sales, as participants were keen to buy plants and supplies to start their own herb gardens.

Building partnerships with local businesses, such as restaurants and grocery stores, can expand your market reach. I established relationships with a few local restaurants that were interested in sourcing fresh, locally-grown herbs and vegetables. These partnerships provided a steady revenue stream and helped promote my greenhouse through their menus and social media channels.

Consider offering loyalty programs or discounts to regular customers. I started a "frequent buyer" program, where customers earned points for every purchase, redeemable for discounts or free products. This not only encouraged repeat business but also helped me track customer preferences and adjust my offerings accordingly.

Legal Considerations

Running a greenhouse business involves navigating a range of legal considerations, from zoning laws to food safety regulations. It might seem daunting at first, but with the right information and preparation, you can ensure your business operates smoothly and within the law.

One of the first steps I took was researching local zoning laws. Not all areas allow commercial agricultural operations, so it's essential to check with your local zoning office to see if your property is zoned for greenhouse operations. In some cases, you might need a special permit or approval from a zoning board. I had to go through a permit process, which involved submitting a business plan and attending a public hearing. While it was a bit nerve-wracking, it ultimately helped solidify my business's presence in the community.

Food safety is another critical area, especially if you're selling edible products. Depending on your location, you may need to comply with regulations set by local health departments or agricultural agencies. These regulations might include guidelines on how you grow, handle, and package your produce. For instance, I had to implement specific cleaning and sanitizing procedures for my greenhouse to meet local food safety standards. It's also important to label your products accurately, particularly if you're selling organic or pesticide-free produce.

If you're planning to sell at farmers' markets or to restaurants, you may need to obtain a business license and register your business with the appropriate state or local authorities. This process often involves choosing a business structure, such as a sole proprietorship, partnership, or corporation, each with its own legal and tax implications. I opted for an LLC (Limited Liability Company) to protect my personal assets and provide flexibility in management.

Insurance is another essential consideration. Liability insurance can protect you in case of accidents or injuries on your property, while crop insurance can safeguard against losses due to weather or pests. I remember the peace of mind I felt after securing liability insurance, knowing that I was covered in case a customer slipped on a wet floor or if a shipment of plants was damaged.

If you're selling plants, particularly if they're rare or protected species, be aware of any regulations regarding plant propagation and sale. Some plants may require permits for cultivation or sale, especially if they are on endangered or protected species lists. I had to navigate these regulations when I started selling certain rare orchids, ensuring I had the necessary permits to propagate and sell them legally.

Intellectual property protection is another area to consider, especially if you're developing unique products or branding. Trademarks can protect your business name, logo, and other branding elements, while patents might be necessary if you've developed a unique growing system or product. I filed for a trademark for my greenhouse's brand name and logo, which helped protect my brand identity and prevented others from using similar branding.

Lastly, keep thorough records of all your business transactions, including sales, expenses, and taxes. Proper bookkeeping not only helps in managing your business effectively but is also crucial during tax season. I hired an accountant to help with my bookkeeping and tax filing, ensuring that I stayed compliant with all tax laws and regulations.

Scaling Up Operations

Scaling up your greenhouse operations is an exciting step, but it requires careful planning and consideration. When I first considered expanding, I was both excited and apprehensive. The idea of growing my business and increasing production was appealing, but I knew it would come with challenges.

One of the first things to consider is whether you have the infrastructure to support expansion. This might involve expanding your physical space, upgrading your greenhouse structures, or investing in new equipment. I started by adding a new section to my greenhouse dedicated to growing leafy greens. This required not only physical expansion but also upgrades to my irrigation and ventilation systems to accommodate the additional plants.

Labor is another critical factor. As your operations grow, so will your need for help. This might mean hiring additional staff or increasing the hours of existing employees. Initially, I handled most of the work myself, but as my greenhouse expanded, I hired a few part-time workers to help with planting, harvesting, and sales. Training them was essential to ensure they understood the specific needs of each plant and the overall operations of the greenhouse.

Marketing and distribution also need to scale with your production. This might involve finding new markets for your products, expanding your online presence, or increasing your participation in farmers' markets. I explored new markets by reaching out to local grocery stores and offering them a consistent supply of fresh produce. Expanding my online store also helped reach customers beyond my local area, offering delivery options for herbs and plants.

Financial planning is crucial when scaling up. This involves budgeting for new equipment, labor costs, marketing expenses, and potential expansions. I sought advice from a business advisor to create a detailed financial plan that accounted for these factors. This planning helped me secure a small business loan to fund the expansion, ensuring I had the capital needed to grow without jeopardizing my existing operations.

Quality control becomes increasingly important as you scale. Maintaining the quality of your produce and plants can be challenging with larger operations, but it's essential for maintaining customer satisfaction and brand reputation. I implemented strict quality control measures, including regular inspections of plants, maintaining detailed records of crop performance, and ensuring consistent care routines across all greenhouse sections.

Customer service should also scale with your operations. As you reach more customers, maintaining a personal touch can be challenging but is crucial for building customer loyalty. I made it a priority to maintain open communication with customers, whether through social media, email newsletters, or face-to-face interactions at markets. This engagement helped build a community around my brand and kept customers coming back.

Lastly, consider the long-term vision for your greenhouse business. Scaling up isn't just about increasing production; it's about building a sustainable and profitable business. I set long-term goals for my greenhouse, including expanding into new product lines, exploring new markets, and continually improving sustainability practices. This vision guided my decisions and helped ensure that every step taken towards expansion aligned with the overall growth strategy.

Scaling up greenhouse operations is a journey filled with challenges and rewards. By planning carefully, maintaining quality, and engaging with your customers, you can successfully expand your operations and enjoy the fruits of your labor on a larger scale.

CHAPTER 21

Harvesting Your Crops, Preserving Your Harvest

"Harvest is not just the end of a growing season; it's the beginning of preserving the fruits of your labor."

In my years of greenhouse gardening, I've found that harvesting and preserving crops is both an art and a science. It's not merely about picking fruits and vegetables; it's about understanding when to harvest, how to handle produce to maintain its quality and the best methods to ensure your hard work pays off long after the growing season has ended.

Harvest Timing and Techniques

Harvesting is one of the most satisfying aspects of greenhouse gardening, but it's important to get it right to ensure the best quality and flavor of your crops. Timing is everything when it comes to harvesting. Each plant has its optimal harvest time, which can vary depending on the type of crop and even the specific variety.

For many crops, the best time to harvest is when they have reached their peak ripeness. For instance, tomatoes should be picked when they're fully colored and slightly firm to the touch. Picking them too early can result in underdeveloped flavor while waiting too long might lead to overripe, mushy fruits. I remember a season when I was overly eager to harvest my tomatoes, only to find that the flavor wasn't as robust as it could have been. Since then, I've learned to carefully monitor the color and firmness, relying on experience and observation to determine the right moment.

Leafy greens like lettuce and spinach are best harvested in the morning when their leaves are crisp and full of moisture. For these, you can use a sharp knife or scissors to cut them just above the base, allowing the plant to regrow for a second harvest. I've found that regularly cutting these greens and allowing them to regrow keeps a continuous supply of fresh leaves.

Root vegetables, such as carrots and beets, need a different approach. They should be harvested when they have reached a desirable size, which you can usually determine by gently brushing away the soil to check their diameter. For carrots, I generally wait until they are at least one inch in diameter, as smaller ones can be quite tough.

When harvesting fruits like cucumbers or peppers, it's essential to use a sharp knife or pruning shears to avoid damaging the plant. Overripe fruits can attract pests and diseases, so it's crucial to harvest them promptly. I use this simple rule: if the fruit is easily detached from the plant with a gentle tug and shows the right color, it's time to harvest.

Each crop will have its specific harvesting guidelines, so be sure to refer to planting guides or resources specific to the crops you're growing. I always keep a detailed journal of my planting and harvesting times, which helps me improve my techniques and understand the unique needs of each crop.

Post-Harvest Handling

Once your crops are harvested, proper handling is essential to maintain their quality and extend their shelf life. Handling crops with care reduces the risk of bruising and damage, which can affect the flavor and longevity of your produce.

Begin by gently brushing off any excess soil from root vegetables and greens. Avoid washing them until you're ready to use or store them, as excess moisture can lead to mold and rot. I remember when I first started gardening, I used to wash everything immediately after harvest. Over time, I learned that keeping produce dry until use or storage helps maintain its freshness.

For fruits and vegetables, such as tomatoes and peppers, it's best to handle them carefully and avoid stacking them too high. Bruising can occur if the weight of the produce presses down on the ones below, leading to spoilage. I use shallow trays or baskets to transport my harvested crops, ensuring that they are well-supported and protected.

Cooling is another crucial step in post-harvest handling. Many crops benefit from being cooled quickly to slow down the ripening process and maintain freshness. For instance, leafy greens should be placed in a cool environment as soon as possible to prevent wilting. I've set up a small refrigeration area in my greenhouse for this purpose, which helps keep greens crisp and fresh.

If you're handling large quantities, such as during a bountiful harvest, it can be helpful to establish a sorting area. This is where you can separate the high-quality produce from any that may be damaged or overripe. Sorting and grading can help ensure that only the best produce makes it to market or into your storage.

For delicate items like herbs, it's essential to handle them gently to avoid bruising. I typically place herbs in breathable containers to allow for air circulation, which helps prevent wilting and maintains their aroma. Proper post-harvest handling sets the foundation for successful preservation and storage.

Storage Methods

Storing your harvested crops properly is key to preserving their freshness and extending their usability. Different crops have unique storage needs, so it's important to understand the best methods for each type.

Root vegetables like carrots and beets can be stored in a cool, dry place, ideally in a root cellar or a similar environment. If you don't have access to a root cellar, a cool basement or an insulated container can work. I use a simple storage setup with wooden crates lined with burlap to keep the vegetables dry and protected from light. Proper ventilation is crucial to prevent mold and rot, so I ensure that my storage area has good airflow.

Leafy greens and herbs, on the other hand, are best stored in the refrigerator. I keep them in crisper drawers or in perforated plastic bags to maintain moisture while allowing for air circulation. This method keeps them fresh for several days. For longer storage, I consider blanching and freezing them to preserve their flavor and nutritional value.

Fruits like apples and pears can be stored in a cool, dark place. I use cardboard boxes lined with newspaper, which helps absorb excess moisture and prevents bruising. Apples should be checked regularly for signs of spoilage, as they can release ethylene gas that accelerates the ripening of other fruits.

For items like tomatoes, which can ripen quickly if not stored properly, I use a single-layer arrangement on countertops or in a cool, dry area. Overripe tomatoes can be used immediately or preserved through canning or freezing.

In general, it's helpful to label your storage containers with the date of harvest to keep track of how long the produce has been stored. This practice helps ensure that you use the oldest items first and prevents spoilage.

Canning and Pickling

Canning and pickling are excellent methods for preserving the abundance of your greenhouse harvest. Both techniques allow you to enjoy your produce long after the growing season has ended.

Canning involves preserving food by sealing it in jars and processing it through heat to kill bacteria and enzymes that cause spoilage. The basic process includes preparing your produce, filling jars with it, and processing them in a water bath or pressure canner. For example, I use canning to preserve tomatoes for sauces and soups. After blanching and peeling the tomatoes, I pack them into sterilized jars with a bit of salt and lemon juice before processing them.

When canning, it's crucial to follow tested recipes and guidelines to ensure safety. Botulism, a rare but serious bacterium, can grow in improperly canned foods, so following guidelines helps prevent contamination. I rely on trusted resources and canning guides to ensure I'm using the correct procedures and processing times for each type of produce.

Pickling, on the other hand, involves preserving food in a solution of vinegar, water, and salt. Pickling not only extends the shelf life of your produce but also adds a tangy flavor. I've had great success pickling cucumbers, beets, and even some herbs. The process is straightforward: prepare your produce, make the pickling brine, and pour it over the produce in sterilized jars. After sealing the jars, they are left to sit for several weeks to develop the pickled flavor.

Both canning and pickling can be customized with various spices and flavorings. I often experiment with different herbs and spices, such as dill, garlic, and mustard seeds, to create unique flavors. These preserved goods make excellent gifts and are a great way to enjoy the fruits of your labor throughout the year.

Drying and Freezing Techniques

Drying and freezing are two more methods for preserving your greenhouse produce, each with its own benefits. Drying removes moisture from food, inhibiting bacterial growth and spoilage. Freezing, on the other hand, preserves food by stopping the growth of microorganisms through low temperatures.

For drying, you can use a food dehydrator, an oven, or even air-dry certain herbs and vegetables. I've used a dehydrator for drying herbs like basil and thyme, which retains their flavor and aroma. Simply wash and cut the herbs, place them on the dehydrator trays, and let the machine do the work. Once dried, store the herbs in airtight containers away from light to preserve their potency.

Fruits like apples, peaches, and tomatoes can also be dried. For these, I slice the fruit thinly and place it on drying racks. The drying process can take several hours to days, depending on the fruit and method used. Dried fruits make great snacks and can be used in baking or cooking.

Freezing is another effective method, especially for preserving large quantities of produce. Before freezing, I blanch vegetables like green beans and corn to stop enzyme action that can affect flavor and texture. Blanching involves briefly boiling the vegetables and then plunging them into ice water. After blanching, I package the vegetables in airtight freezer bags or containers and label them with the date.

For fruits, I wash and cut them into pieces before freezing. I often pre-freeze them on a baking sheet to prevent clumping, then transfer them to bags or containers. Freezing preserves the flavor and texture of fruits and vegetables, making it easy to enjoy them in smoothies, soups, and stews.

Both drying and freezing require proper storage to maintain quality. For dried produce, keep it in a cool, dark place in airtight containers. For frozen items, ensure your freezer is set to 0°F (-18°C) and avoid overloading it, as proper air circulation is essential for maintaining consistent temperatures.

Making Jams, Sauces, and Preserves

Making jams, sauces, and preserves is a delightful way to enjoy and share your greenhouse harvest. These methods not only extend the shelf life of your produce but also offer a chance to create delicious, homemade products.

Jams and jellies are made by cooking fruit with sugar and pectin to create a sweet, spreadable product. For a simple strawberry jam, I mash fresh strawberries and cook them with sugar and pectin until the mixture reaches the desired consistency. Once cooled, the jam is poured into sterilized jars and sealed. Homemade jams make great gifts and add a personal touch to breakfast or dessert.

Sauces, such as tomato sauce or salsa, can be made using ripe tomatoes and various seasonings. I cook down tomatoes with garlic, onions, and herbs to create a rich, flavorful sauce. The sauce can be canned for long-term storage or frozen for convenience. Making your own sauces allows you to control the ingredients and customize the flavors to your liking.

Preserves involve cooking fruits with sugar and sometimes spices to create a chunky, sweet spread. I've made peach preserves by cooking diced peaches with sugar and lemon juice until thickened. The preserves are then poured into jars and sealed. Preserves retain the texture of the fruit and can be enjoyed on toast, in desserts, or as a topping for yogurt.

Each of these preservation methods has its own charm and benefits. Experimenting with different recipes and techniques can lead to a bounty of homemade goodies that showcase your greenhouse harvest. These preserved items not only enhance your culinary repertoire but also provide a way to savor your crops long after the growing season has ended.

CONCLUSION

Congratulations! You've made it through this comprehensive guide on greenhouse gardening, and I couldn't be more thrilled to have shared this journey with you. Whether you're a seasoned gardener or just starting out, you now have a wealth of knowledge at your fingertips to help you make the most of your greenhouse.

We've covered a lot of ground, and as we wrap up, let's take a moment to recap the key points and celebrate the progress we've made.

Summary and Key Points

- **Understanding Greenhouse Climate**

We began our journey by delving into the importance of understanding and managing greenhouse climate. I emphasized the need for precise temperature control, humidity management, and light requirements. Monitoring climate conditions and adjusting for seasonal changes are crucial for creating an optimal environment for plant growth. The tools for climate control that we discussed will help you maintain the right conditions throughout the year, ensuring your plants thrive.

- **Ventilation and Air Circulation**

Proper ventilation and air circulation are vital for a healthy greenhouse environment. We explored the significance of airflow and ventilation systems, including the differences between passive and active ventilation. I provided tips on installation and maintenance, helping you ensure that your greenhouse remains well-ventilated and free of stagnant air, which can lead to issues like fungal growth and poor plant health.

- **Heating and Cooling Your Greenhouse**

Maintaining optimal temperatures in your greenhouse is essential for plant growth, especially during extreme weather conditions. We discussed various heating options, cooling techniques, and insulation strategies. By understanding how to manage seasonal temperature fluctuations and applying energy-efficient practices, you can keep your greenhouse environment stable and supportive for your plants.

- **Setting Up Your Greenhouse for Success**

Setting up your greenhouse correctly from the start lays a strong foundation for successful gardening. We covered choosing the right growing medium, selecting containers, and essential gardening tools. Preparing the greenhouse interior and organizing planting areas are crucial steps, and the initial plant setup ensures that your plants have the best possible start.

- **Selecting Plants for Your Greenhouse**

Choosing the right plants for your greenhouse is a key factor in maximizing your yield and enjoyment. We discussed various factors to consider, including popular greenhouse plants and companion planting strategies. By selecting appropriate seeds and seedlings and following seasonal planting guides, you can create a thriving and diverse plant collection in your greenhouse.

- **Seed Starting and Propagation Techniques**

Starting plants from seeds and using propagation methods are techniques that allow you to grow a wide variety of plants efficiently. We explored different propagation methods and how to utilize your greenhouse for this purpose. Care for young plants, timing, scheduling, and troubleshooting common issues were all covered, giving you a solid understanding of how to nurture new growth from seed to maturity.

- **Planting and Growing Techniques in a Greenhouse**

We dove into the specific techniques for planting and growing in a greenhouse environment. Proper planting techniques, watering requirements, fertilization, pest control, and disease prevention are all essential components of successful greenhouse gardening. Regular pruning and maintenance ensure that your plants remain healthy and productive throughout their growing cycle.

- **Maintaining Your Greenhouse Throughout the Year**

Maintaining your greenhouse year-round is key to keeping it in top condition. We discussed seasonal maintenance tasks, preparation for different seasons, and troubleshooting common issues. Regular cleaning and upkeep, along with equipment maintenance, are vital for ensuring that your greenhouse continues to support healthy plant growth.

- **Hydroponics for Beginners**

Hydroponics offers a soilless approach to growing plants, which can be highly efficient and productive. We covered the basics of hydroponic systems, including maintaining nutrient solutions and setting up your system. Choosing hydroponic plants and understanding system maintenance are crucial for successful hydroponic gardening.

- **Aquaponics: Integrating Fish and Plants**

Aquaponics combines aquaculture and hydroponics to create a symbiotic environment where fish and plants support each other's growth. We explored the concept of aquaponics, setting up an aquaponic system, and maintaining water quality. Choosing the right fish and plants, harvesting, and balancing the system is key to successful aquaponic gardening.

- **Greenhouse Automation**

Automating greenhouse tasks can significantly enhance efficiency and productivity. We discussed various types of automation systems, setting up and monitoring these systems, and adjusting them to meet your needs. The benefits of automation include increased precision and ease of management, and we also touched on future trends in greenhouse technology.

- **Extending Your Growing Season**

Extending your growing season allows you to maximize harvests and enjoy your greenhouse produce for longer. Techniques for using season extenders, selecting short-growing varieties, and managing light and temperature were explored. Proper timing and protection against frost are also essential for maintaining a productive greenhouse throughout the year.

- **Sustainable Practices for Your Greenhouse**

Sustainability is a crucial aspect of modern greenhouse gardening. We covered practices for reducing environmental impact, organic pest control, water conservation, and using sustainable building materials. Implementing energy-efficient practices and recycling and waste management strategies will help you create an eco-friendly greenhouse operation.

- **Expanding Your Greenhouse Operations**

As your greenhouse grows, expanding your operations can open up new opportunities. We discussed diversifying your crops, using the greenhouse for non-food plants, and turning it into a business. Marketing your greenhouse products and understanding legal considerations are important for scaling up operations effectively.

- **Harvesting Your Crops, Preserving Your Harvest**

Finally, harvesting and preserving your crops allows you to enjoy the fruits of your labor long after the growing season ends. We covered techniques for harvesting, post-harvest handling, and various storage methods. Canning, pickling, drying, freezing, and making jams and preserves were all discussed as ways to extend the enjoyment of your greenhouse bounty.

As we conclude this journey through greenhouse gardening, I want to congratulate you on your dedication and hard work. You've now equipped yourself with a comprehensive understanding of greenhouse operations, from the basics to advanced techniques. You've learned how to manage climate conditions, ventilate effectively, and maintain optimal temperatures. You've explored methods for seed starting, plant care, and extending your growing season, and you've delved into sustainable practices and innovative systems like hydroponics and aquaponics.

Gardening in a greenhouse is a rewarding experience that allows you to grow a diverse range of plants in a controlled environment. It offers the opportunity to experiment, innovate, and continuously improve your techniques. Remember, every greenhouse is unique, and your journey will be shaped by your specific conditions and goals. Keep track of your experiences, learn from each season, and never stop experimenting with new methods and ideas.

Your greenhouse is more than just a space for growing plants; it's a reflection of your passion and commitment to gardening. Whether you're growing for personal enjoyment or turning your greenhouse into a business, the skills and knowledge you've gained will serve you well. Embrace the challenges, celebrate the successes, and always strive to learn and grow.

Thank you for joining me on this adventure. I hope you feel inspired and empowered to create a thriving greenhouse that brings you joy and productivity. Here's to many successful seasons ahead, full of lush plants, abundant harvests, and the satisfaction of nurturing life from seed to harvest. Happy gardening!

More from the Author:

Take Your Gardening to the Next Level with Beekeeping

As a reader of *Greenhouse Gardening Made Easy for Beginners*, you already know the joy of nurturing plants and watching them thrive. Now, imagine enriching your garden's growth and vitality by introducing bees to your space. These incredible pollinators not only boost the health and productivity of your plants but also provide the golden reward of fresh, home-harvested honey.

To guide you in starting and managing a successful bee colony, we present *The Ultimate Beekeeping Guide: Build, Manage, and Profit From Your Own Bee Colony*. Whether you're new to beekeeping or an experienced gardener, this book offers clear, practical steps to seamlessly combine gardening with beekeeping.

Highlights of the Book:

- **Step-by-Step Guidance:** From choosing the right bees to harvesting your first batch of honey, we've got you covered.
- **Expert Insights:** Practical tips for managing a thriving hive with ease.
- **Profit Opportunities:** Learn how beekeeping can be both a rewarding hobby and a source of income.

Bring new life to your garden, support biodiversity, and enjoy the sweet rewards of beekeeping. Order your copy of *The Ultimate Beekeeping Guide* today and take the first step toward this fulfilling endeavor in your favorite store.

ISBN:

Ebook: 978-1965739068

Paperback: 978-1965739075

REFERENCES

Hessayon, D.G. (1994). The Greenhouse Expert. Expert Books.

Ball, V. (1997). Ball RedBook: Greenhouse Growing. Ball Publishing.

Bradley, S. (2017). The Complete Guide to Greenhouse Gardening. Lorenz Books.

Jabbour, N. (2011). The Year-Round Vegetable Gardener. Storey Publishing.

Neal, C. D. (1975). Build Your Own Greenhouse: How to Construct, Equip, and Maintain It. UNKNO publishers

Smith, S. (2000) Greenhouse Gardener's Companion: Growing Food & Flowers in Your Greenhouse or Sunspace. Fulcrum Publishing.

Gillman, J. H. (2008) The Truth About Organic Gardening: Benefits, Drawbacks, and the Bottom Line. Timber Press.

Mefferd, A. (2017). The Greenhouse and Hoophouse Grower's Handbook: Organic Vegetable Production Using Protected Culture. Chelsea Green Publishing.

Hartmann, H., Kester, D., Davies, F., Geneve, R., & Wilson, S. (2017). Plant Propagation: Principles and Practices. Pearson Publishing.

Johns, J. (2016). Greenhouse Gardening for Beginners - A Beginners Guide To Growing Fruit and Vegetables All Year Round. CreateSpace Independent Publishing Platform.

Jones, H. G. (2014). Plants and Microclimate: A Quantitative Approach to Environmental Plant Physiology. Cambridge University Press.

Jones, J. B. (2012). Complete Guide for Growing Plants Hydroponically. CRC press.

Baras, T. (2018). DIY Hydroponic Gardens: How to Design and Build an Inexpensive System for Growing Plants in Water. Cool Springs Press.

Resh, H. (1987). Hydroponic Food Production: A Definitive Guidebook for the Advanced Home Gardener and the Commercial Hydroponic Grower. CRC Press.

Celine, W. (2016). Aquaponics: How to Build Your Own Aquaponic System. CreateSpace Independent Publishing Platform.

Stone, M. (2014). Greenhouse Gardening: How to Build and Maintain a Greenhouse Garden. CreateSpace Independent Publishing Platform.

Styer, R. C., & Koranski, D. S. (2011). Plug and Transplant Production: A Grower's Guide. Ball Publishing.

Marshall, R. (2014). The Greenhouse Gardener's Manual. Timber Press.

Williams, J. (2013). Greenhouse Gardening 101: How To Build a Greenhouse For Food, Flowers and Fun. Vertical Gardening Group.

Greenhouse Hacks & Tips | What I Learned Year 1 | joegardener®

Royal Horticultural Society. (2021). Vegetables: growing in your greenhouse / RHS Gardening

Coleman, E. (2009). The Winter Harvest Handbook: Year-Round Vegetable Production Using Deep-Organic Techniques and Unheated Greenhouses. Chelsea Green Publishing.

Stone, C. (2015). The Urban Farmer: Growing Food for Profit on Leased and Borrowed Land. New Society Publishers.

Made in the USA
Monee, IL
04 November 2026

67378711R00174